Current Directions in COGNITIVE SCIENCE

EDITED BY

Barbara A. Spellman
University of Virginia

Daniel T. Willingham
University of Virginia

Upper Saddle River, New Jersey 07458

Upper Saddle River, New Jersey 07458

1010 Vermont Avenue, NW
Suite 1100
Washington, D.C. 20005-4907

10 9 8 7 6

ISBN 0-13-191991-1

Printed in the United States of America

Contents

Readings from
Current Directions in Psychological Science

Overview of the Book

Picking articles for this reader was, in many ways, a pleasure. As we looked back through the years of *Current Directions* to find articles relevant to cognitive science, we were struck by how many were interesting, well-written, or both. The hard part then was how to narrow down the selection to the 20-25 for the reader.

We assumed that most users of this book would be concurrently taking a course in cognitive psychology or cognitive science. Thus, we decided not to pick articles that were on issues likely to be well-covered in the textbooks for those courses. This decision also led to another decision—not to pick articles that were more than 6 years old. For one thing, such topics were likely to already be in the textbooks; for another, the reader is titled *Current* (not *Historic*) *Directions.*

Beyond that, the reader reflects our shared views as to topics that are particularly timely in cognitive psychology and topics that are applicable beyond cognitive psychology. Indeed, over the past years we have each had reason to answer the question: What are the current trends in cognitive psychology and where do you see the field going? In sum we would say that the directions are micro, macro, and down under. By "micro" we mean the current interest in neuroscience: What is the relation between mind and brain? As the tools for looking at the brain get better, we can learn more about how the mind works. But as Miller and Keller (2000) point out, don't be fooled, one is not reducible to the other. By "macro" we mean two things: first, ties to systems bigger than our individual minds (e.g., to evolution and culture) and second, to human affairs such as law, business, and economics, whose practioners increasingly look to cognitive scientists for information about how people think and behave in the "real world" outside the laboratory. And by "down under" we mean that interest in the unconscious and the role of emotion and affect in influencing cognition is booming. We find all of these directions very exciting and have tried to select articles that illustrate them and, we hope, bring the excitement to you.

Textbooks are good for explaining the theories in a field and for detailing a few central research findings. These articles give you a chance to read recent reviews by experts that will show you something beyond the basics, by taking you to the cutting edge of the field. We hope that these readings will expand your understanding of what cognitive science is and can do. Enjoy.

Visual Perception

Cognitive psychology can be frustrating because it seems so splintered. Beginning students study perception, then attention, then working memory, and so on. They have the sense that each researcher has fenced off a small domain of the mind for study, and scarcely cares or even acknowledges that the mind performs other functions, let alone that these functions might have some influence on the researcher's cherished domain.

There is a grain of truth in this perception. Cognitive psychologists do believe that progress can be made by examining cognitive processes in isolation, because they believe that it is useful to consider the mind as *modular*. That means that the mind is composed of processes, each of which performs a particular cognitive function (e.g., calculating the motion of objects, or maintaining auditory information in working memory). The modules are not entirely independent of one another—they communicate—but they have a certain independence in that their function does not radically change based on what other modules are doing. In addition, most introductory cognitive psychology books don't discuss interaction because it is a more advanced topic—you need to know the basics before discussing how the basics interact. This section includes three articles that consider the interaction of visual perception with other cognitive processes: attention, imagery, and other perceptual systems.

Our intuition tells us that attention increases the processing of visual information, but vision still proceeds to some extent without the benefit of attention. For example, suppose you had walked past a painting in a museum but your friend told you that it was her favorite, so you went back and inspected it more closely. You would likely say that you had *seen* the painting when you first passed it, but had not noticed many of the details until you had directed attention to it. This observation indicates that attention enables deeper or more complex visual perception, but that some perceptual processes operate independently of attention. These processes allow you to know that it's a painting on a wall, and not a mirror or a window. Research reviewed by Mack in "Inattentional Blindness: Looking Without Seeing" suggests that our intuition is not quite right, and that visual perception is more dependent on attention than we would guess. It appears that we do not see objects to which we don't attend. But the interesting twist is that this generalization holds true only for conscious perception. Visible objects can and do affect behavior, although we may not be aware of how they influence us.

Attention relates to perception by making certain types of perception possible. Perception has an altogether different relationship to visual imagery. Again using our intuition as a guide, we might guess that the two have something in common. When asked "does the Statue of Liberty hold her torch in her right hand or her left hand?" most people report that they answer this question by generating a mental picture of the Statue and then inspecting it. This mental picture seems to have something in common with perception—it entails a visual experience—but further reflection reveals important differences between perception and visual imagery. Pause from your reading for a moment and image a tiger. Once you have done so, try to count the number of stripes on the tiger's torso. Most people find this task impossible, although you could of course complete the task if you were perceiving a tiger. Images are less detailed than percepts. Is imagery a watered-down version of perception? No, but it is now known that there is considerable overlap in the two systems. In "The Mind's Eye Mapped onto the Brain's Matter", Behrmann describes in detail the current knowledge of what these systems have in common, and how they differ.

It is easy for us to appreciate how vision and mental imagery overlap; it is more difficult to imagine significant overlap of vision and other sensory systems, such as audition. One might describe the tone of Stan Getz's saxophone as "warm" or even refer to it as "golden" but we mean that as a metaphor, not literally. "Synesthesia: Strong and Weak" by Martino and Marks reports on the small number of people for whom such descriptions are not metaphoric. Synesthetes experience cross talk among the senses such that stimulation of one sensory modality (e.g., a sound) leads to a strong sensation in another modality (e.g., a visual image). Although true synesthesia is rare, it can inform us about basic mechanisms of how perceptions are coded.

The articles in this section illustrate the intricate relationships among cognitive processes. Although we may speak of processes such as "attention" as though they operate in isolation, cognitive psychologists are mindful that this is a convenient simplification, and that even as we study individual cognitive processes we must bear in mind how they interact.

Inattentional Blindness: Looking Without Seeing

Arien Mack[1]
Psychology Department, New School University, New York, New York

Abstract

Surprising as it may seem, research shows that we rarely see what we are looking at unless our attention is directed to it. This phenomenon can have serious life-and-death consequences. Although the inextricable link between perceiving and attending was noted long ago by Aristotle, this phenomenon, now called inattentional blindness (IB), only recently has been named and carefully studied. Among the many questions that have been raised about IB are questions about the fate of the clearly visible, yet unseen stimuli, whether any stimuli reliably capture attention, and, if so, what they have in common. Finally, is IB an instance of rapid forgetting, or is it a failure to perceive?

Keywords

inattention; perception; awareness

Imagine an experienced pilot attempting to land an airplane on a busy runway. He pays close attention to his display console, carefully watching the airspeed indicator on his windshield to make sure he does not stall, yet he never sees that another airplane is blocking his runway!

Intuitively, one might think (and hope) that an attentive pilot would notice the airplane in time. However, in a study by Haines (1991), a few experienced pilots training in flight simulators proceeded with their landing when a clearly visible airplane was blocking the runway, unaware of the second airplane until it was too late to avoid a collision.

As it turns out, such events are not uncommon and even may account for many car accidents resulting from distraction and inattention. This is why talking on cell telephones while driving is a distinctly bad idea. However, the pervasive assumption that the eye functions like a camera and our subjective impression of a coherent and richly detailed world lead most of us to assume that we see what there is to be seen by merely opening our eyes and looking. Perhaps this is why we are so astonished by events like the airplane scenario, although less potentially damaging instances occur every day, such as when we pass by a friend without seeing her.

These scenarios are examples of what psychologists call inattentional blindness (IB; Mack & Rock, 1998). IB denotes the failure to see highly visible objects we may be looking at directly when our attention is elsewhere. Although IB is a visual phenomenon, it has auditory and tactile counterparts as well; for example, we often do not hear something said to us if we are "not listening."

INATTENTIONAL BLINDNESS

The idea that we miss a substantial amount of the visual world at any given time is startling even though evidence for such selective seeing was first reported in the 1970s by Neisser (1979). In one of several experiments, he asked participants to view a video of two superimposed ball-passing games in which one group of players wore white uniforms and another group wore black uniforms. Participants counted the number of passes between members of one of the groups. When the participants were subsequently asked to report what they had seen, only 21% reported the presence of a woman who had unexpectedly strolled though the basketball court carrying an open umbrella, even though she was clearly in view some of the time. Researchers recently replicated this finding with a study in which a man dressed in a gorilla costume stopped to thump his chest while walking through the court and remained visible for between 5 and 9 s (Simons & Chabris, 1999).

Although it is possible that some failures to see the gorilla or the umbrella-carrying woman might have resulted from not looking directly at them, another body of work supports the alternative explanation that the observers were so intent on counting ball passes that they missed the unexpected object that appeared in plain view. Research I have conducted with my colleagues (Mack & Rock, 1998) conclusively demonstrates that, with rare exceptions, observers generally do not see what they are looking directly at when they are attending to something else. In many of these experiments, observers fixated on specified locations while simultaneously attending to a demanding perceptual task, the object of which might be elsewhere. Under these conditions, observers often failed to perceive a clearly visible stimulus that was located exactly where they were looking.

INATTENTIONAL BLINDNESS OR INATTENTIONAL AMNESIA?

Not surprisingly, there is a controversy over whether the types of failures documented in such experiments are really evidence that the observers did not see the stimulus, or whether they in fact saw the stimulus but then quickly forgot it. In other words, is IB more correctly described as *inattentional amnesia* (Wolfe, 1999)? Although this controversy may not lend itself to an empirical resolution, many researchers find it difficult to believe that a thumping gorilla appearing in the midst of a ball game is noticed and then immediately forgotten. What makes the argument for inattentional amnesia even more difficult to sustain is evidence that unseen stimuli are capable of priming, that is, of affecting some subsequent act. (For example, if a subject is shown some object too quickly to identify it and is then shown it again so that it is clearly visible, the subject is likely to identify it more quickly than if it had not been previously flashed. This is evidence of priming: The first exposure speeded the response to the second.) Priming can occur only if there is some memory of the stimulus, even if that memory is inaccessible.

UNCONSCIOUS PERCEPTION

A considerable amount of research has investigated unconscious, or *implicit*, perception and those perceptual processes that occur outside of awareness. This work has led many researchers to conclude that events in the environment, even if not consciously perceived, may direct later behavior. If stimuli not seen because of IB are in fact processed but encoded outside of awareness, then it should be possible to demonstrate that they prime subsequent behavior.

The typical method for documenting implicit perception entails measuring reaction time over multiple trials. Such studies are based on the assumption that an implicitly perceived stimulus will either speed up or retard subsequent responses to relevant stimuli depending on whether the priming produces facilitation or inhibition.[2] However, because subjects in IB experiments cannot be made aware of the critical stimulus, unlike in many kinds of priming studies, only one trial with that stimulus is possible. This requirement rules out reaction time procedures, which demand hundreds of trials because reaction time differences tend to be small and therefore require stable response rates that can be achieved only with many trials. Fortunately, an alternate procedure, stem completion, can be used when the critical stimuli are words. In this method, some observers (experimental group) are exposed to a word in an IB procedure, and other observers (control group) are not. Then, the initial few letters of the unseen word are presented to all the observers, who are asked to complete the string of letters with one or two English words. If the members of the experimental group complete the string with the unseen word more frequently than do the members of the control group, this is taken as evidence that the experimental group implicitly perceived and encoded the word.

IB experiments using this method have demonstrated significant priming (Mack & Rock, 1998), as well as other kinds of evidence that visual information undergoes substantial processing prior to the engagement of attention. For example, evidence that aspects of visual processing take place before attention is allocated has been provided by a series of ingenious IB experiments by Moore and her collaborators (e.g., Moore & Egeth, 1997). This work has shown that under conditions of inattention, basic perceptual processes, such as those responsible for the grouping of elements in the visual field into objects, are carried out and influence task responses even though observers are unable to report seeing the percepts that result from those processes. For example, in one study using a modification of the IB procedure, Moore and Egeth investigated the Müller-Lyer illusion, in which two lines of equal length look unequal because one has outgoing fins, which make it look longer, and the other has ingoing fins, which make it look shorter. In this case, the fins were formed by the grouping of background dots: Dots forming the fins were closer together than the other dots in the background. Moore and Egeth demonstrated that subjects saw the illusion even when, because of inattention, the fins were not consciously perceived. Whatever processes priming entails, the fact that it occurs is evidence of implicit perception and the encoding of a stimulus in memory. Thus, the fact that the critical stimulus in the IB paradigm can prime subsequent responses is evidence that this stimulus is implicitly perceived and encoded.

When Do Stimuli Capture Attention and Why?

That unconsciously perceived stimuli in IB experiments undergo substantial processing in the brain is also supported by evidence that the select few stimuli able to capture attention when attention is elsewhere are complex and meaningful (e.g., the observer's name, an iconic image of a happy face) rather than simple features like color or motion. This fact suggests that attention is captured only after the meaning of a stimulus has been analyzed. There are psychologists who believe that attention operates much earlier in the processing of sensory input, before meaning has been analyzed (e.g., Treisman, 1969). These accounts, however, do not easily explain why modest changes, such as inverting a happy face and changing one internal letter in the observer's name, which alter the apparent meaning of the stimuli but not their overall shape, cause a very large increase in IB (Mack & Rock, 1998).

Meaning and the Capture of Attention

If meaning is what captures attention, then it follows axiomatically that meaning must be analyzed before attention is captured, which is thought to occur at the end stage of the processing of sensory input. This therefore implies that even those stimuli that we are not intending to see and that do not capture our attention must be fully processed by the brain, for otherwise their meanings would be lost before they had a chance of capturing our attention and being perceived. If this is the case, then we are left with some yet-unanswered, very difficult questions. Are all the innumerable stimuli imaged on our retinas really processed for meaning and encoded into memory, and if not, which stimuli are and which are not?

Although we do not yet have answers to these questions, an unpublished doctoral dissertation by Silverman, at New School University, has demonstrated that there can be priming by more than one element in a multielement display, even when these elements cannot be reported by the subject. This finding is relevant to the question whether all elements in the visual field are processed and stored because up to now there has been scarcely any evidence of priming by more than one unreportable element in the field. The fact of multielement priming begins to suggest that unattended or unseen elements are processed and stored, although it says nothing about how many elements are processed and whether the meaning of all the elements is analyzed.

One answer to the question of how much of what is not seen is encoded into memory comes from an account of perceptual processing based on the assumption that perception is a limited-capacity process and that processing is mandatory up to the point that this capacity is exhausted (Lavie, 1995). According to this analysis, the extent to which unattended objects are processed is a function of the difficulty of the perceptual task (i.e., the perceptual load). When the perceptual load is high, only attended stimuli are encoded. When it is low, unattended stimuli are also processed. This account faces some difficulty because it is not clear how perceptual load should be estimated. Beyond this, however, it is difficult to reconcile this account with evidence suggesting that observers are likely to see their own names even when they occur among the stimuli that must be ignored in order to perform a demanding perceptual task (Mack, Pappas, Silverman, & Gay, 2002). It should be noted, however, that these latter results are at odds with a published report (Rees, Russell, Firth, & Driver, 1999) I describe in the next section.

EVIDENCE FROM NEURAL IMAGING

Researchers have used magnetic imaging techniques to try to determine what happens in the brain when observers fail to detect a visual stimulus because their attention is elsewhere. Neural recording techniques may be able to show whether visual stimuli that are unconsciously perceived arouse the same areas of the brain to the same extent as visual stimuli that are seen. This is an important question because it bears directly on the nature of the processing that occurs outside of awareness prior to the engagement of attention and on the difference between the processing of attended and unattended stimuli.

In one study, Scholte, Spekreijse, and Lamme (2001) found similar neural activity related to the segregation of unattended target stimuli from their backgrounds (i.e., the grouping of the unattended stimuli so they stood out from the background on which they appeared), an operation that is thought to occur early in the processing of visual input. This activation was found regardless of whether the stimuli were attended and seen or unattended and not seen, although there was increased activation for targets that were attended and seen. This finding is consistent with the behavioral findings of Moore and Egeth (1997), cited earlier, showing that unattended, unseen stimuli undergo lower-level processing such as grouping, although the additional neural activity associated with awareness suggests that there may be important differences in processing of attended versus unattended stimuli.

In another study, Rees and his colleagues (Rees et al., 1999) used functional magnetic resonance imaging (fMRI) to picture brain activity while observers were engaged in a perceptual task. They found no evidence of any difference between the neural processing of meaningful and meaningless lexical stimuli when they were ignored, although when the same stimuli were attended to and seen, the neural processing of meaningful and meaningless stimuli did differ. These results suggest that unattended stimuli are not processed for meaning. However, in another study that repeated the procedure used by Rees et al. (without fMRI recordings) but included the subject's own name among the ignored stimuli, many subjects saw their names, suggesting that meaning was in fact analyzed (Mack et al., 2002). Thus, one study shows that ignored stimuli are not semantically processed, and the other suggests that they are. This conflict remains unresolved. Are unattended, unseen words deeply processed outside of awareness, despite these fMRI results, which show no evidence of semantic neural activation by ignored words? How can one reconcile behavioral evidence of priming by lexical stimuli under conditions of inattention (Mack & Rock, 1998) with evidence that these stimuli are not semantically processed?

NEUROLOGICAL DISORDER RELATED TO INATTENTIONAL BLINDNESS

People who have experienced brain injuries that cause lesions in the parietal cortex (an area of the brain associated with attention) often exhibit what is called unilateral visual neglect, meaning that they fail to see objects located in the visual field opposite the site of the lesion. That is, for example, if the lesion is on the right, they fail to eat food on the left side of their plates or to shave the left half of their faces. Because these lesions do not cause any sensory deficits, the apparent blindness cannot be attributed to sensory causes and has been explained in terms of the role of the parietal cortex in attentional processing

(Rafal, 1998). Visual neglect therefore seems to share important similarities with IB. Both phenomena are attributed to inattention, and there is evidence that in both visual neglect (Rafal, 1998) and IB, unseen stimuli are capable of priming. In IB and visual neglect, the failure to see objects shares a common cause, namely inattention, even though in one case the inattention is produced by brain damage, and in the other the inattention is produced by the task. Thus, evidence of priming by neglected stimuli appears to be additional evidence of the processing and encoding of unattended stimuli.

ATTENTION AND PERCEPTION

IB highlights the intimate link between perception and attention, which is further underscored by recent evidence showing that unattended stimuli that share features with task-relevant stimuli are less likely to suffer IB than those that do not (Most et al., 2001). This new evidence illustrates the power of our intentions in determining what we see and what we do not.

CONCLUDING REMARKS

Although the phenomenon of IB is now well established, it remains surrounded by many unanswered questions. In addition to the almost completely unexplored question concerning whether all unattended, unseen stimuli in a complex scene are fully processed outside of awareness (and if not, which are and which are not), there is the question of whether the observer can locate where in the visual field the information extracted from a single unseen stimulus came from, despite the fact that the observer has failed to perceive it. This possibility is suggested by the proposal that there are two separate visual systems, one dedicated to action, which does not entail consciousness, and the other dedicated to perception, which does entail consciousness (Milner & Goodale, 1995). That is, the action stream may process an unseen stimulus, including its location information, although the perception stream does not. An answer to this question would be informative about the fate of the unseen stimuli.

The pervasiveness of IB raises another unresolved question. Given that people see much less than they think they do, is the visual world a mere illusion? According to one provocative answer to this question, most recently defended by O'Regan and Noe (2001), the outcome of perceptual processing is not the construction of some internal representation; rather, seeing is a way of exploring the environment, and the outside world serves as its own external representation, eliminating the need for internal representations. Whether or not this account turns out to be viable, the phenomenon of IB has raised a host of questions, the answers to which promise to change scientists' understanding of the nature of perception. The phenomenon itself points to the serious dangers of inattention.

Recommended Reading

Mack, A., & Rock, I. (1998). (See References)
Rensink, R. (2002). Change blindness. *Annual Review of Psychology*, 53, 245–277.

Simons, D. (2000). Current approaches to change blindness. *Visual Cognition, 7*, 1–15.
Wilkens, P. (Ed.). (2000). Symposium on Mack and Rock's *Inattentional Blindness. Psyche*, 6 and 7. Retrieved from http://psyche.cs. monash.edu.au/psyche-indexv7.html//ib

Acknowledgments—I am grateful for the comments and suggestions of Bill Prinzmetal and Michael Silverman.

Notes

1. Address correspondence to Arien Mack, Psychology Department, New School University, 65 Fifth Ave., New York, NY 10003.

2. An example of a speeded-up response (facilitation, or positive priming) has already been given. Negative, or inhibition, priming occurs when a stimulus that has been actively ignored is subsequently presented. For example, if a series of superimposed red and green shapes is rapidly presented and subjects are asked to report a feature of the red shapes, later on it is likely to take them longer to identify the green shapes than a shape that has not previously appeared, suggesting that the mental representation of the green shapes has been associated with something like an "ignore me" tag.

References

Haines, R.F. (1991). A breakdown in simultaneous information processing. In G. Obrecht & L.W. Stark (Eds.), *Presbyopia research* (pp. 171–175). New York: Plenum Press.
Lavie, N. (1995). Perceptual load as a necessary condition for selective attention. *Journal of Experimental Psychology: Human Perception and Performance, 21*, 451–468.
Mack, A., Pappas, Z., Silverman, M., & Gay, R. (2002). What we see: Inattention and the capture of attention by meaning. *Consciousness and Cognition, 11*, 488–506.
Mack, A., & Rock, I. (1998). *Inattentional blindness.* Cambridge, MA: MIT Press.
Milner, D., & Goodale, M.A. (1995). *The visual brain in action.* Oxford, England: Oxford University Press.
Moore, C.M., & Egeth, H. (1997). Perception without attention: Evidence of grouping under conditions of inattention. *Journal of Experimental Psychology: Human Perception and Performance, 23*, 339–352.
Most, S.B., Simons, D.J., Scholl, B.J., Jimenez, R., Clifford, E., & Chabris, C.F. (2001). How not to be seen: The contribution of similarity and selective ignoring to sustained inattentional blindness. *Psychological Science, 12*, 9–17.
Neisser, U. (1979). The control of information pickup in selective looking. In A.D. Pick (Ed.), *Perception and its development: A tribute to Eleanor Gibson* (pp. 201–219). Hillsdale, NJ: Erlbaum.
O'Regan, K., & Noe, A. (2001). A sensorimotor account of vision and visual consciousness. *Behavioral and Brain Sciences, 25*, 5.
Rafal, R. (1998). Neglect. In R. Parasuraman (Ed.), *The attentive brain* (pp. 489–526). Cambridge, MA: MIT Press.
Rees, G., Russell, C., Firth, C., & Driver, J. (1999). Inattentional blindness versus inattentional amnesia. *Science, 286*, 849–860.
Scholte, H.S., Spekreijse, H., & Lamme, V.A. (2001). Neural correlates of global scene segmentation are present during inattentional blindness [Abstract]. *Journal of Vision, 1*(3), Article 346. Retrieved from http://journalofvision. org/1/3/346
Simons, D.J., & Chabris, C.F. (1999). Gorillas in our midst: Sustained inattentional blindness for dynamic events. *Perception, 28*, 1059–1074.
Treisman, A. (1969). Strategies and models of selective attention. *Psychological Review, 76*, 282–299.
Wolfe, J. (1999). Inattentional amnesia. In V. Coltheart (Ed.), *Fleeting memories* (pp. 71–94). Cambridge, MA: MIT Press.

Critical Thinking Questions

1. This article notes that some traffic accidents may well be caused by inattentional blindness, but you have likely had the experience of daydreaming as you drove without incident. In fact, your inattention may be so complete that you might drive to a familiar destination (e.g., your home) and feel surprised when you arrive. Does this phenomenon mean that the low-level processing that occurs with inattention is sufficient to support successful driving?
2. This article describes evidence that priming is supported by low-level processes that occur even in the absence of attention. When attention is directed to stimuli, you are aware of them. Other than awareness, what cognitive processes do you think are possible only with attention?
3. Mack cites the neurological phenomenon of unilateral neglect as sharing important similarities with inattentional blindness. Neglect patients fail to process stimuli in *all* modalities, not just vision. Do you think that lack of attention yields "blindness" in other modalities? In other words, if you are not attending, will you not hear an auditory stimulus? Will you not feel a tactile stimulus such as someone touching you?

The Mind's Eye Mapped Onto the Brain's Matter

Marlene Behrmann[1]

Department of Psychology, Carnegie Mellon University, Pittsburgh, Pennsylvania

Abstract

Research on visual mental imagery has been fueled recently by the development of new behavioral and neuroscientific techniques. This review focuses on two major topics in light of these developments. The first concerns the extent to which visual mental imagery and visual perception share common psychological and neural mechanisms: although the research findings largely support convergence between these two processes, there are data that qualify the degree of overlap between them. The second issue involves the neural substrate mediating the process of imagery generation. The data suggest a slight left-hemisphere advantage for this process, although there is considerable variability across and within subjects. There also remain many unanswered questions in this field, including what the relationship is between imagery and working memory and what representational differences, if any, exist between imagery and perception.

Keywords

mental imagery; visual perception; cognitive neuroscience

Consider sitting in your office and answering the question "How many windows do you have in your living room?" To decide how to answer, you might construct an internal visual representation of your living room from the stored information you possess about your home, inspect this image so as to locate the windows, and then count them. This type of internal visual representation (or "seeing with the mind's eye"), derived in the absence of retinal stimulation, is known as visual mental imagery, and is thought to be engaged in a range of cognitive tasks including learning, reasoning, problem solving, and language. Although much of the research on mental imagery has been concerned specifically with visual mental imagery, and hence the scope of this review is restricted to this topic, similar internal mental representations exist in the auditory and tactile modalities and in the motor domain.

The past decade has witnessed considerable progress in our understanding of the psychological and neural mechanisms underlying mental imagery. This is particularly dramatic because, in the not-too-distant past, during the heyday of behaviorism, discussions of mental imagery were almost banished from scientific discourse: Given that there was no obvious way of measuring so private an event as a mental image and there was no homunculus available for viewing the pictures in the head even if they did exist, the study of mental imagery fell into disrepute. Indeed, through the 1940s and 1950s, Psychological Abstracts recorded only five references to imagery. The study of mental imagery was revived in the 1970s, through advances such as the experiments of Shepard and colleagues (Shepard & Cooper, 1982) and of

Kosslyn and colleagues (see Kosslyn, 1994), and the dual coding theory of Paivio (1979).

Although a general consensus endorsing the existence of mental imagery began to emerge, it was still not fully accepted as a legitimate cognitive process. Some researchers queried whether subjects were simply carrying out simulations of their internal representations in symbolic, nonvisual ways rather than using a visual, spatially organized code. Other researchers suggested that subjects were simply conforming to the experimenters' expectations and that the data that appeared to support a visual-array format for mental imagery merely reflected the experimenters' belief in this format rather than the true outcome of a mental imagery process (Pylyshyn, 1981). In recent years, powerful behavioral and neuroscientific techniques have largely put these controversies to rest. This review presents some of this recent work.

RELATIONSHIP BETWEEN VISUAL IMAGERY AND VISUAL PERCEPTION

Perhaps the most hotly debated issue is whether mental imagery exploits the same underlying mechanisms as visual perception. If so, generating a visual mental image might be roughly conceived of as running perception backward. In perception, an external stimulus delivered to the eye activates visual areas of the brain, and is mapped onto a long-term representation that captures some of the critical and invariant properties of the stimulus. During mental imagery, the same long-term representations of the visual appearance of an object are used to activate earlier representations in a top-down fashion through the influence of preexisting knowledge. This bidirectional flow of information is mediated by direct connections between higher-level visual areas (more anterior areas dealing with more abstract information) and lower-level visual areas (more posterior areas with representations closer to the input).

Mental Imagery and Perception Involve Spatially Organized Representations

Rather than being based on propositional or symbolic representations, mental images appear to embody spatial layout and topography, as does visual perception. For example, many experiments have shown that the distance that a subject travels in mental imagery is equivalent to that traveled in perceptual performance (e.g., imagining the distance between New York and Los Angeles vs. looking at a real map to judge the distance). Recent neuroimaging studies have also provided support for the involvement of spatially organized representations in visual imagery (see Kosslyn et al., 1999); for example, when subjects form a high-resolution, depictive mental image, primary and secondary visual areas of the occipital lobe (areas 17 and 18, also known as V1 and V2), which are spatially organized, are activated.[2] Additionally, when subjects perform imagery, larger images activate relatively more anterior parts of the visual areas of the brain than smaller images, a finding consistent with the known mapping of how visual information from the world is mediated by different areas of primary visual cortex. Moreover, when repetitive transcranial magnetic stimulation[3] is applied and disrupts the normal function of area 17, response times in both perceptual and imagery tasks increase, further supporting the involvement of primary visual areas in mental imagery (Kosslyn et al., 1999).

Shared Visual and Imagery Areas Revealed Through Functional Imaging

Not only early visual areas but also more anterior cortical areas can be activated by imagined stimuli; for example, when subjects imagine previously seen motion stimuli (such as moving dots or rotating gratings), area MT/MST, which is motion sensitive during perception, is activated (Goebel, Khorram-Sefat, Muckli, Hacker, & Singer, 1998). Color perception and imagery also appear to involve some (but not all) overlapping cortical regions (Howard et al., 1998), and areas of the brain that are selectively activated during the perception of faces or places are also activated during imagery of the same stimuli (O'Craven & Kanwisher, in press). Higher-level areas involved in spatial perception, including a bilateral parieto-occipital network, are activated during spatial mental imagery, and areas involved in navigation are activated during mental simulation of previously learned routes (Ghaem et al., 1997). As is evident, there is considerable overlap in neural mechanisms implicated in imagery and in perception both at lower and at higher levels of the visual processing pathways.

Neuropsychological Data for Common Systems

There is also neuropsychological evidence supporting the shared-systems view. For example, many patients with cortical blindness (i.e., blindness due to damage to primary visual areas of the brain) or with scotomas (blind spots) due to destruction of the occipital lobe have an associated loss of imagery, and many patients with visual agnosia (a deficit in recognizing objects) have parallel imagery deficits. Interestingly, in some of these latter cases, the imagery and perception deficits are both restricted to a particular domain; for example, there are patients who are unable both to perceive and to image only faces and colors, only facial emotions, only spatial relations, only object shapes and colors, or only living things. The equivalence between imagery and perception is also noted in patients who, for example, fail both to report and to image information on the left side of space following damage to the right parietal lobe.

There are, however, also reports of patients who have a selective deficit in either imagery or perception. This segregation of function is consistent with the functional imaging studies showing that roughly two thirds of visual areas of the brain are activated during both imagery and perception (Kosslyn, Thompson, & Alpert, 1997). That is, selective deficits in imagery or perception may be explained as arising from damage to the nonoverlapping regions. Selective deficits are particularly informative and might suggest what constitutes the nonoverlapping regions. Unfortunately, because the lesions in the neuropsychological patients are rather large, one cannot determine precise anatomical areas for these nonoverlapping regions, but insights into the behaviors selectively associated with imagery or perception have been obtained, as discussed next.

Patients with impaired imagery but intact perception are unable to draw or describe objects from memory, to dream, or to verify propositions based on memory ("does the letter W have three strokes?"). It has been suggested that in these cases, the process of imagery generation (which does not overlap with perception) may be selectively affected without any adverse consequences for recognition. I review this generation process in further detail in a later section. It has also been suggested that low- or intermediate-level processes may play a greater (but not exclusive) role in perception than they do in imagery. For example, when asked to image a "kangaroo," one accesses the intact long-term representation of a kangaroo, and this is then instantiated and available for inspection. When one is perceiving a kangaroo, however, featural analysis, as well as perceptual organization such as figure-ground segregation and feature grouping or integration, is required. If a patient has a perceptual deficit because of damage to these low- or intermediate-level processes, the patient will be unable to perceive the display, but imagery might well be spared because it relies less on these very processes (Behrmann, Moscovitch, & Winocur, 1994).

Summary and Challenges

There is substantial evidence that imagery and perception share many (although not all) psychological and neural mechanisms. However, some results suggest that the early visual regions in the occipital lobe are not part of the shared network. For example, it has been shown that reliable occipital activation is not always observed in neuroimaging studies when subjects carry out mental imagery tasks. A possible explanation for these null results concerns the task demands: When the task does not require that the subjects form a highly depictive image, no occipital activation is obtained. A second explanation might have to do with subject sampling: There are considerable individual differences in mental imagery ability, and subjects who score poorly on a mental imagery vividness questionnaire show less blood flow in area 17 than those who score higher. If a study includes only low-imagery individuals, no occipital lobe activation might be obtained. A final explanation concerns the nature of the baseline, or control, condition used in neuroimaging experiments: If subjects are instructed to rest but instead continue to activate internal representations, when the activation obtained in the baseline is subtracted from that obtained in the imagery condition, no primary visual cortex activation will be observed.

Another challenge to the conclusion that primary visual areas are involved in imagery comes from studies of patients who have bilateral occipital lesions and complete cortical blindness but preserved imagery. Indeed, some of these patients are capable of generating such vivid images that they believe these to be veridical perceptions. For example, when a set of keys was held up before one such subject, she correctly identified the stimulus based on the auditory signal but then went on to provide an elaborate visual description of the keys, convinced that she could actually see them (Goldenberg, Müllbacher, & Nowak, 1995). Another subject with bilateral occipital lesions whose perception was so impaired that he could not even differentiate light from dark was still able to draw well from memory and performed exceptionally well on standard imagery tasks (Chatterjee & Southwood, 1995).

In sum, although the data supporting a strong association between imagery and perception are compelling, some findings are not entirely consistent with this conclusion.

GENERATION OF MENTAL IMAGES

A second major debate in the imagery literature concerns the mechanisms involved in generating a mental image. This is often assumed to be a process specific to imagery (or perhaps more involved in imagery than in perception) and involves the active construction of a long-term mental representation. Although there has been debate concerning whether there is such a process at all, several lines of evidence appear to support its existence and role in imagery. There are, for example, reports of neuropsychological patients who have preserved perception but impaired imagery and whose deficit is attributed to image generation. Patient R.M., for example, could copy well and make good shape discriminations of visually presented objects but could not draw even simple shapes nor complete from memory visually presented shapes that were partially complete (Farah, Levine, & Calvanio, 1988).

One controversial and unanswered issue concerns the neural substrate of the generation process. The growing neuropsychological literature has confirmed the preponderance of imagery generation deficits in patients with lesions affecting the left temporo-occipital lobe regions. There is not, however, a perfect relationship between this brain region and imagery generation, as many patients with such lesions do not have an impairment in imagery generation.

In many studies, normal subjects show a left-hemisphere advantage for imagery generation; when asked to image half of an object, subjects are more likely to image the right half, reflecting greater left-hemisphere than right-hemisphere participation in imagery generation. Additionally, right-handed subjects show a greater decrement in tapping with their right than left hand while performing a concurrent imagery task, reflecting the interference encountered by the left hemisphere while tapping and imaging simultaneously. (The left hemisphere controls movement on the right side of the body.) Studies in which information is presented selectively to one visual field (and thereby one hemisphere) have, however, yielded more variable results with normal subjects. Some studies support a left-hemisphere superiority, some support a right-hemisphere superiority, and some find no hemispheric differences at all. Studies with split-brain patients[4] also reveal a trend toward left-hemisphere involvement, but also some variability. Across a set of these rather rare patients, imagery performance is better when the stimulus is presented to the left than to the right hemisphere, although this finding does not hold for every experiment and the results are somewhat variable even within a single subject.

Neuroimaging studies in normal subjects have also provided some support for a left-hemisphere basis for imagery generation. For example, functional magnetic resonance imaging showed more activation of the left inferior occipital-temporal region when subjects generated images of heard words compared with when they were simply listening to these words (D'Esposito et al., 1997). This result had also been observed previously using ERPs[5]; an asymmetry in the waveforms of the two hemispheres implicated the left temporo-occipital regions in imagery generation (see Farah, 1999).

In sum, there is a slight but not overwhelming preponderance of evidence favoring the left hemisphere as mediating the imagery generation process. A conservative conclusion from these studies suggests that there may well be some degree of left-hemisphere specialization, but that many individuals have some capability for imagery generation by the right-hemisphere. Another suggestion is that both hemispheres are capable of imagery generation, albeit in different ways; for example, subjects showed a left-hemisphere advantage in a generation task when they memorized how the parts of a stimulus were arranged but showed a right-hemisphere advantage when they memorized the metric positions of the parts and how they could be "mentally glued" together (see Kosslyn, 1994).

CONCLUSION

Although considerable progress has been made in analyzing the convergence between imagery and perception, there are several outstanding issues. One of these is the relationship between imagery and the activation of internal representations in other cognitive tasks. For example, during visual working memory tasks, a mental representation of an object or spatial location is maintained over a delay period, in the absence of retinal stimulation. In these tasks, areas in the very front of the brain, rather than occipital cortex, are activated despite the similarities between this task and mental imagery. Similarly, in tasks that involve top-down forms of attention, subjects are verbally instructed to search for a target (such as a red triangle) in an upcoming display. Although subjects likely generate an image of a red triangle, this is generally not conceived of as an instance of mental imagery, and in neuro imaging studies the location of activation is not usually sought in occipital cortex.

Another perplexing and unresolved issue concerns the reasons that vivid imagery and hallucinations are not confused with reality, especially given that functional imaging studies show identical activations during hallucinations and perception (ffytche et al., 1998). Several solutions to this dilemma have been proposed, among them the idea that hallucinations derive from a failure to self-monitor an inner voice, with the result that the source of the stimulus is located in the external world. A second explanation suggests that perceptions are deeper and contain more detail than images. As Hume (1739/1963) stated, "The difference betwixt these [imagery-ideas and perception] consists in the degree of force and liveliness, with which they strike upon the mind . . . . Perceptions enter with most force and violence . . . . By ideas I mean the faint images of these in thinking and reasoning" (p. 311). How to verify these claims empirically is not obvious, yet this issue clearly demands resolution.

Recommended Reading

Behrmann, M., Moscovitch, M., & Winocur, G. (1999). Visual mental imagery. In G.W. Humphreys (Ed.), *Case studies in vision* (pp. 81–110). London: Psychology Press.
Farah, M.J. (1999). (See References)
Kosslyn, S.M. (1994). (See References)
Richardson, J.T.E. (1999). *Imagery*. Philadelphia: Psychology Press.

Acknowledgments—This work was supported by grants from the National Institute of Mental Health (MH54246 and MH54766). I thank Martha Farah and Steven Kosslyn for helpful discussions about mental imagery and Nancy Kanwisher for her insightful comments on this manuscript.

Notes

1. Address correspondence to Marlene Behrmann, Department of Psychology, Carnegie Mellon University, Pittsburgh, PA 15213-3890; e-mail: behrmann+@cmu.edu.

2. Visual information is received initially via the retina of the eye and is then transmitted through various visual pathways to the brain. This information is sent initially to the primary visual area of the brain, housed posteriorly in the occipital cortex, and is then sent more anteriorly through secondary visual areas to the temporal lobe of the brain for the purposes of recognition. The primary visual area is also known as area 17 or V1, and the secondary area is known as area 18 or V2. The visual input is also sent from the occipital cortex up to the parietal areas of the brain, which represent and code spatial information.

3. Repetitive transcranial magnetic stimulation is a new method in which magnetic pulses are delivered to the brain from a magnet placed externally on the scalp. The electrical pulses disrupt the function of the underlying brain area temporarily and are thus analogous to a reversible lesion. This method allows investigators to determine the involvement of certain brain areas in particular cognitive processes.

4. Split-brain patients are individuals who have undergone a separation of the two sides of the brain (cerebral hemispheres). This is done in individuals who have intractable and uncontrolled epilepsy in order to prevent the seizure activity from spreading across the entire brain. Unfortunately, it also prevents the transfer of all other forms of information from one hemisphere to the other.

5. ERPs, or evoked response potentials, are recordings of the brain's electrical activity in response to stimuli that are presented to the subject. The potentials are measured over time as waveforms obtained from electrodes placed at specific sites on the scalp, and different waveforms roughly reflect differential participation of some brain sites in the task under examination.

References

Behrmann, M., Moscovitch, M., & Winocur, G. (1994). Intact visual imagery and impaired visual perception in a patient with visual agnosia. *Journal of Experimental Psychology: Human Perception and Performance, 20*, 1068–1087.

Chatterjee, A., & Southwood, M.H. (1995). Cortical blindness and visual imagery. *Neurology, 45*, 2189–2195.

D'Esposito, M., Detre, J.A., Aguirre, G.K., Stallcup, M., Alsop, D.C., Tippett, L.J., & Farah, M.J. (1997). Functional MRI study of mental image generation. *Neuropsychologia, 35*, 725–730.

Farah, M.J. (1999). Mental imagery. In M. Gazzaniga (Ed.), *The cognitive neurosciences* (Vol. 2, pp. 965–974). Cambridge, MA: MIT Press.

Farah, M.J., Levine, D.N., & Calvanio, R. (1988). A case study of a mental imagery deficit. *Brain and Cognition, 8*, 147–164.

ffytche, D.H., Howard, R.J., Brammer, M.J., David, A., Woodruff, P., & Williams, S. (1998). The anatomy of conscious vision: An fMRI study of visual hallucinations. *Nature Neuroscience, 1*, 738–742.

Ghaem, O., Mellet, E., Crivello, F., Tzourio, N., Mazoyer, B., Berthoz, A., & Denis, M. (1997). Mental navigation along memorized routes activates the hippocampus, precuneus and insula. *NeuroReport, 8*, 739–744.

Goebel, R., Khorram-Sefat, D., Muckli, L., Hacker, H., & Singer, W. (1998). The constructive nature of vision: Direct evidence from functional magnetic resonance imaging studies of apparent motion and motion imagery. *European Journal of Neuroscience, 10*, 1563–1573.
Goldenberg, G., Müllbacher, W., & Nowak, A. (1995). Imagery without perception—A case study of anosognosia for cortical blindness. *Neuropsychologia, 33*, 39–48.
Howard, R.J., ffytche, D.H., Barnes, J., McKeefry, D., Ha, Y., Woodruff, P.W., Bullmore, E.T., Simmons, A., Williams, S.C.R., David, A.S., & Brammer, M. (1998). The functional anatomy of imagining and perceiving colour. *NeuroReport, 9*, 1019–1023.
Hume, D. (1963). A treatise of human nature. In V.C. Chappel (Ed.), *The philosophy of David Hume* (pp. 11–311). New York: Modern Library. (Original work published 1739)
Kosslyn, S.M. (1994). *Image and brain*. Cambridge, MA: MIT Press.
Kosslyn, S.M., Pascual-Leone, A., Felician, O., Camposano, S., Keenan, J.P., Thompson, W.L., Ganis, G., Sukel, K.E., & Alpert, N.M. (1999). The role of Area 17 in visual imagery: Convergent evidence from PET and TMS. *Science, 284*, 167–170.
Kosslyn, S.M., Thompson, W.L., & Alpert, N.M. (1997). Neural systems shared by visual imagery and visual perception: A positron emission tomography study. *Neuroimage, 6*, 320–334.
O'Craven, K., & Kanwisher, N. (in press). Mental imagery of faces and places activates corresponding stimulus-specific brain regions. *Journal of Cognitive Neuroscience*.
Paivio, A. (1979). *Imagery and verbal processes*. Hillsdale, NJ: Erlbaum.
Pylyshyn, Z.W. (1981). The imagery debate: Analogue media versus tacit knowledge. *Psychological Review, 88*, 16–45.
Shepard, R.N., & Cooper, L.A. (1982). *Mental images and their transformations*. Cambridge, MA: MIT Press.

Critical Thinking Questions

1. There are individual differences in imagery capabilities. Some people are slow or inaccurate in answering questions that require imagery (e.g., mentally rotating objects) whereas others are fast. If the mechanisms for imagery and perception overlap, would you expect parallel individual differences in perception?

2. If imagery and perception share mechanisms, one might expect that it would be difficult to use both functions simultaneously. For example, one might expect that subjects would close their eyes to reduce perceptual interference when performing an imagery task. Can you think of evidence from your own experience that such interference might take place?

3. How do dreams compare to images and to perception? Dreams seem more like perception because both can be quite vivid, but mental images usually are not. Also, imagery requires constant attention to be maintained, whereas dreams do not. On the other hand, dreams can be likened to imagery because neither entails retinal stimulation. Do you think that dreams are more like imagery or more like perception? Why?

Synesthesia: Strong and Weak

Gail Martino[1] and Lawrence E. Marks

The John B. Pierce Laboratory, New Haven, Connecticut (G.M., L.E.M.), and Department of Diagnostic Radiology (G.M.) and Department of Epidemiology and Public Health (L.E.M.), Yale Medical School, Yale University, New Haven, Connecticut

Abstract

In this review, we distinguish strong and weak forms of synesthesia. Strong synesthesia is characterized by a vivid image in one sensory modality in response to stimulation in another one. Weak synesthesia is characterized by cross-sensory correspondences expressed through language, perceptual similarity, and perceptual interactions during information processing. Despite important phenomenological dissimilarities between strong and weak synesthesia, we maintain that the two forms draw on similar underlying mechanisms. The study of strong and weak synesthetic phenomena provides an opportunity to enrich scientists' understanding of basic mechanisms involved in perceptual coding and cross-modal information processing.

Keywords

synesthesia; cross-modal perception; selective attention

Color is central to Carol's life. As a professional artist, she uses color to create visual impressions in her paintings. Yet unlike most people, Carol also uses color to diagnose her health. She is able to accomplish this by consulting the colored images she sees in connection with pain. For example, a couple of years ago, Carol fell and damaged her leg badly while climbing on rocks at the beach. She diagnosed the severity of her accident not only by the intensity of her pain, but also by the intensity of the orange color that spread across her mind's eye. She said, "When I saw that everything was orange, I knew I should be rushed to the hospital."

Carol's tendency to see colors in response to pain is an example of strong synesthesia. Synesthesia means "to perceive together," and strong synesthesia occurs when a stimulus produces not only the sensory quality typically associated with that modality, but also a quality typically associated with another modality. Strong synesthesia typically arises on its own, although it also can follow the ingestion of drugs such as mescaline and LSD. In this article, we confine our discussion to synesthesia unrelated to drug use.

Over the two centuries since strong synesthesia was first identified in the scientific literature, several heterogeneous phenomena have been labeled as synesthetic. These phenomena range from strong experiences like Carol's, on the one hand, to weaker cross-modal literary expressions, on the other. We believe it is a mistake to label all of these phenomena simply as synesthesia because the underlying mechanisms cannot be identical, although they may overlap. In this review, we distinguish between *strong synesthesia*, which describes the unusual experiences of individuals such as Carol, and *weak synesthesia*, which describes milder forms of cross-sensory connections revealed through language and perception. In both types of synesthesia, cross-modal correspondences are evident, suggesting that the neural processes underlying strong and weak synesthesia, although not wholly identical, nonetheless may have a common core.

STRONG SYNESTHESIA

The Synesthete

Strong synesthesia is an uncommon condition with an unusual demographic profile. Although estimates vary, one recent estimate places the incidence at 1 in 2,000, with females outnumbering males 6 to 1 (Baron-Cohen, Burt, Laittan-Smith, Harrison, & Bolton, 1996). Strong synesthesia clusters in families, leading some researchers to suggest that it has a genetic basis (Baron-Cohen et al., 1996). Empirical evidence for this notion remains sparse.

Except for the finding that there are more female than male synesthetes, few other generalizations characterize strong synesthetes as a group. Attempts have been made to link synesthesia with artistic creativity: Several strong synesthetes described in case studies have worked in the visual arts or music (Cytowic, 1989). Furthermore, several artists who have produced highly creative work—Kandinsky, Rimbaud, and Scriabin—have drawn inspiration from synesthesia. It is unlikely, however, that these artists were themselves strong synesthetes (Dann, 1998). Thus, there are no empirical data to support the idea that strong synesthetes show high artistic creativity.

Strong Synesthetic Correspondence

Case studies offer several characteristics of strong synesthesia (see Cytowic, 1989). In all cases, an association or correspondence exists between an *inducer* in one modality (e.g., pain in Carol's case) and an *induced* percept or image in another (e.g., color). These correspondences have several salient characteristics. For example, they can be idiosyncratic and systematic at the same time. Correspondences are idiosyncratic in that each synesthete has a unique scheme of associations. Middle C on the piano may be blue to one color-music synesthete and green to another. Yet both synesthetes will reveal a systematic relationship between color brightness or lightness and auditory pitch: The higher the pitch of the sound, the lighter or brighter the color of the image (Marks, 1978). Thus C', an octave above middle C, will evoke a correspondingly brighter blue or a lighter green synesthetic color. Besides pitch-brightness and pitch-lightness, auditory-visual synesthesia reveals several other systematic associations. Notable is the association of pitch to shape and size: The higher the pitch of the sound, the sharper, more angular, and smaller the visual image (Marks, 1978).

Synesthetic images are typically simple (e.g., consist of a color or shape), but dynamic (e.g., as the inducer waxes and wanes, so does the image). In some cases, the induced image is so vivid as to be distracting. S, a mnemonist (professional memorizer) and synesthete described by Luria (1968), explained that "crumbly and yellow" images coming from a speaker's mouth were so intense that

S had difficulty attending to the intended message. This observation exemplifies another general principle—induced images tend to be visual, whereas inducing stimuli tend to be auditory, tactile, or gustatory (Cytowic, 1989). The reason for this asymmetry is unknown.

Because strong synesthesias are noticed in early childhood, it is possible they are inborn. Many strong synesthetes claim that their cross-modal experiences "have always been there" (e.g., Cytowic, 1989). Harrison and Baron-Cohen (1997) argued that the higher incidence of synesthesia in females speaks against synesthesia being learned. If synesthesia is learned, why should more synesthetes be female than male?

The connection between the inducer and induced is so entrenched that the image is considered part of the percept's literal identity. For S to refer to a voice as "crumbly and yellow" is to offer a literal rather than a metaphorical description. Given the close connection between the inducer and the induced, one would expect correspondences to be highly memorable and durable. In a study of color-word synesthesia (Baron-Cohen et al., 1996), strong synesthetes and control subjects were asked to report the color induced by many words (the control subjects named the first color that came to mind). One hour later, the strong synesthetes were 97% accurate in recalling pairs, whereas the control subjects were 13% accurate. Baron-Cohen et al. argued that this result means that synesthetic perception is highly memorable and genuine. It is not clear to us, however, whether to attribute the synesthetes' superior performance to their synesthesia or perhaps to better memory for word-color pairings in general.

Processing

Reports of strong synesthetes offer clues about how strong synesthesia is produced and where it arises. With regard to production, the relation between the inducer and the induced is typically unidirectional. That is, although a voice induces a yellow image, a yellow percept need not induce an image of a voice.

Some investigators postulate that strong synesthesia arises from a disorder within low-level sensory mechanisms. The sensory leakage hypothesis claims that information leaks from one sensory channel into another, producing strong synesthesia (Harrison & Baron-Cohen, 1997). Leakage might occur, for example, if nerve cells fail to form or migrate properly during neonatal development.

Another hypothesis places the mechanism within specific regions of the brain. Measurement of cerebral blood flow in a single strong synesthete, M.W., suggests that increased activation of areas involved in memory and emotion (i.e., limbic areas) and simultaneous suppression of areas involved in higher reasoning (i.e., cortical areas) produce synesthetic perceptions (Cytowic, 1989). This "limbic-cortex disconnection" has not yet been replicated (see Frith & Paulesu, 1997). Failures to replicate may be related to the unique nature of strong synesthesia in each individual or due to methodological differences across studies.

Besides knowing where strong synesthesia occurs, it is important to know how and why it takes on its phenomenological form (meaning known through the senses, rather than through thought or intuition). In this regard, case studies of strong synesthetes have provided some insights. Yet case studies are not sufficient to

Table 1 Summary of claims about strong and weak synesthesia

Characteristic	Synesthesia	
	Strong	Weak
Prevalance	Uncommon, gender bias favoring females	Common
Experience of pairings	One stimulus is perceived, the other is experienced as an image	Both stimuli are perceived
Organization of correspondences	Idiosyncratic and systematic	Systematic
Definition of correspondences	Aboslute	Contextual
Role of learning	Some may be unlearned	Some learned, some unlearned
Semantic association	Literal	Metaphorical
Memory	Easily identified and remembered	Easily identified and remembered
Processing	Unidirectional at a low-level sensory locus	Bidirectional at a high-level semantic locus

explain how and why strong synesthetes' perceptions differ from the norm. Toward this end, future research should be guided by a perceptual or cognitive framework. Table 1 offers a summary of characteristics such a framework must address.

WEAK SYNESTHESIA

The phenomenology of strong synesthesia led us to ask whether individuals who lack strong synesthesia nevertheless show analogous cross-modal associations. There is considerable evidence that one can create, identify, and appreciate cross-modal connections or associations even if one is not strongly synesthetic. These abilities constitute weak synesthesia. One form of association is the cross-modal metaphor found in common language (e.g., *warm* color and *sweet* smell) and in literature (e.g., Baudelaire's poem "Correspondences"). Other evidence for weak synesthesia comes from the domain of music. Some people believe that music, like language, contains cross-modal connections. For example, the idea that pitches and colors are associated motivated the invention of the color organ by Castel and inspired the composition *Prometheus* by Scriabin.

Laboratory experiments square with the idea that most people can appreciate cross-modal associations. In such studies, participants are asked to pair a stimulus from one sensory modality to a stimulus from another. These studies show that pairings are systematic. For example, given a set of notes varying in pitch and a set of colors varying in lightness, the higher the pitch, the lighter the color paired with it (see Marks, 1978). This pitch-lightness relation resembles the one observed in strong synesthesia, with one notable difference. In weak synesthesia, the correspondences are defined by context, so that the highest pitch is always associated with the lightest color. Here lies a distinction between strong and weak synesthesia: Although cross-modal correspondences in weak synesthesia are systematic and contextual, those in strong synesthesia are systematic and absolute (display a one-to-one mapping). Despite this difference, it appears that strong and weak synesthetes share an understanding of how visual and auditory dimensions are related (Marks, 1978).

Role of Learning

Are cross-modal correspondences inborn or learned? The answer appears to be, a bit of both. Infants who have not yet learned language show a kind of cross-modal "matching" of loudness-brightness (Lewkowicz & Turkewitz, 1980) and pitch-position (Wagner, Winner, Cicchetti, & Gardner, 1981). Other correspondences develop over time. For example, 4-year-old childrencan match pitch and brightness systematically, but not pitch and visual size. By age 12, children perform these matches as well as do adults (Marks, Hammeal, & Bornstein, 1987).

Processing

What mental processes underlie the ability to form cross-modal associations? The self-reports of strong synesthetes indicate cross-modal interactions are unidirectional and may involve sensory processes. We wondered whether these characteristics of strong synesthesia apply to cross-modal processing more generally. To address this issue, we developed a cross-modal selective attention task. This task measures a person's ability to respond to a stimulus in one modality while receiving concurrent input from a different, "unattended" modality. If an unattended stimulus affects your ability to respond to an attended one, then the two stimuli are said to interact during information processing.

In a typical task, participants may be asked to classify a sound (high or low tone) in the presence of a color (black or white square), so there are four possible combinations of sounds and colors. Participants are faster at classifying high-pitched tones when these are accompanied by white (vs. black) colors, and are faster at identifying low-pitched tones when they are accompanied by black (vs. white) colors. Analogous results are obtained when participants classify the lightness of a square and color is the unattended stimulus (Melara, 1989). This pattern of findings is termed a congruence effect. Congruence effects entail superior performance when attended and unattended stimuli "match" cross-modally (e.g., high pitch + white square; low pitch + black square) rather than "mismatch" (e.g., high pitch + black square; low pitch + white square). Congruence effects suggest that (a) there is cross-modal interaction (unattended signals can affect one's ability to make decisions about attended ones), (b) the cross-modal correspondence between stimuli is important in determining when interactions occur, and (c) interactions are bidirectional (congruence effects can occur when either sounds or colors are attended). (See Martino & Marks, 1999, for converging evidence.)

Why do congruence effects occur? Two accounts predominate. According to *a sensory hypothesis*, congruence effects involve absolute correspondences processed within low-level sensory mechanisms. These correspondences may arise from common properties in underlying neural codes (e.g., temporal properties of neural impulses may link visual brightness to auditory pitch). This account is consistent with the sensory leakage theory of strong synesthesia and with reports that infants show cross-modal correspondences. Alternatively, we propose that congruence effects involve high-level mechanisms, which develop over childhood from experience with percepts and language—an idea we term the *semantic-coding hypothesis* (SCH; e.g., Martino & Marks, 1999).

The SCH makes four claims. First, although cross-modal correspondence may arise from sensory mechanisms in infants, these correspondences reflect postsensory (meaning-based) mechanisms in adults. Second, experience with percepts from various modalities and the language a person uses to describe these percepts produces an abstract semantic network that captures synesthetic correspondence. Third, when synesthetically corresponding stimuli are perceived, they are recoded from sensory representations into abstract ones based on this semantic network. Fourth, the coding of stimuli from different modalities as matching or mismatching depends on the context within which the stimuli are presented. As mentioned previously, cross-modal matches are defined contextually, so the stimuli are perceived as matching or mismatching only when two or more values are presented in each modality.

Critical support for the SCH comes from selective attention studies like the one described earlier. For example, congruence effects occur when tones and colors both vary from trial to trial, but not when either the tone or the color remains constant (Melara, 1989). The sensory account incorrectly predicts that matches should be processed more efficiently than mismatches in both conditions because correspondences are absolute. The SCH explains the result as a context effect: Trial-by-trial variation provides a context in which to define stimulus values relative to one another, thus highlighting a synesthetic association.

Even stronger evidence for the SCH is that linguistic stimuli are sufficient to drive congruence effects. That is, like the colors black and white, the words *black* and *white* produce congruence effects when paired with low- and high-pitched tones (see Martino & Marks, 1999). The sensory hypothesis cannot account for these findings because interaction is claimed to occur at a sensory level. The SCH accounts for them by proposing that all stimuli (sensory and linguistic) are recoded postperceptually into a single abstract representation that captures the synesthetic correspondence between them.

CONCLUSIONS

Synesthesia is not a unitary phenomenon, but instead takes on strong and weak forms. Strong and weak synesthesia differ in phenomenology, prevalence, and perhaps even some mechanisms underlying their expression. Whereas strong synesthesia expresses itself in perceptual experience proper, weak synesthesia is most clearly evident in cross-modal metaphorical language and in cross-modal matching and selective attention.

Several questions about the nature of strong and weak synesthesia await further investigation. Some concern cognitive and neurological underpinnings: Is strong synesthesia mediated by semantic codes, as weak synesthesia appears to be? Are the brain regions involved in the two kinds of synesthesia similar? For example, do both recruit sensory and semantic areas of the brain? Other issues concern development: To what extent may strong synesthesia be learned or embellished over time? The opportunity to tackle such fundamental questions makes synesthesia an exciting topic for future research.

Recommended Reading

Dann, K.T. (1998). (See References) Harrison, J.E., & Baron-Cohen, S. (1997). (See References)

Marks, L. (1978). (See References)

Acknowledgments—Support was provided by National Institutes of Health (NIH) Training Grant T32 DC00025-13 to the first author and by NIH Grant R01 DC02752 to the second author.

Note

1. Address correspondence to Gail Martino, The John B. Pierce Laboratory, 290 Congress Ave., New Haven, CT 06519; e-mail: gmartino@jbpierce.org.

References

Baron-Cohen, S., Burt, L., Laittan-Smith, F., Harrison, J.E., & Bolton, P. (1996). Synaesthesia: Prevalence and familiarity. *Perception*, 25, 1073–1080.

Cytowic, R.E. (1989). *Synaesthesia: A union of the senses.* Berlin: Springer.

Dann, K.T. (1998). *Bright colors falsely seen.* New Haven, CT: Yale University Press.

Frith, C.D., & Paulesu, E. (1997). The physiological basis of synaesthesia. In S. Baron-Cohen & J.E. Harrison (Eds.), *Synaesthesia: Classic and contemporary readings* (pp. 123–147). Cambridge, MA: Blackwell.

Harrison, J.E., & Baron-Cohen, S. (1997). Synaesthesia: A review of psychological theories. In S. Baron-Cohen & J.E. Harrison (Eds.), *Synaesthesia: Classic and contemporary readings* (pp. 109–122). Cambridge, MA: Blackwell.

Lewkowicz, D.J., & Turkewitz, G. (1980). Cross-modal equivalence in early infancy: Auditory-visual intensity matching. *Developmental Psychology*, *16*, 597–601.

Luria, A.R. (1968). *The mind of a mnemonist.* New York: Basic Books.

Marks, L. (1978). The unity of the senses: Interrelations among the modalities. New York: Academic Press.

Marks, L.E., Hammeal, R.J., & Bornstein, M.H. (1987). Perceiving similarity and comprehending metaphor. *Monographs of the Society for Research in Child Development*, 52(*1*, Serial No. 215).

Martino, G., & Marks, L.E. (1999). Perceptual and linguistic interactions in speeded classification: Tests of the semantic coding hypothesis. *Perception*, 28, 903–923.

Melara, R.D. (1989). Dimensional interactions between color and pitch. *Journal of Experimental Psychology: Human Perception and Performance*, 15, 69–79.

Wagner, S., Winner, E., Cicchetti, D., & Gardner, H. (1981). "Metaphorical" mapping in human infants. *Child Development*, 52, 728–731.

Critical Thinking Questions

1. How could one use brain imaging data to test the sensory leakage and the semantic coding hypotheses?
2. Is the semantic coding hypothesis for weak synesthesia just another way of describing metaphor? For example, suppose I demonstrate that most people

agree on a mapping between concepts where one of them is *not* sensory (e.g., honesty is warm, but revenge is cold). Does this mean that weak synesthesia is just metaphor applied to sensation?

3. Synesthesia is described as a mapping between different sensations. Could there be a mapping between movement and sensation (e.g., a synesthete experiences a particular color when a particular hand movement is made). If we don't observe that sort of mapping, does its lack tell us anything interesting about synesthesia?

Memory

Memory is usually the most extensively covered topic in Cognitive Psychology textbooks. It is a central cognitive function and is customarily placed in the middle of those books. Without memory, what use would perception be when we would not recognize what we have seen from one instance to the next? Without memory, how good would reasoning and judgment be when we could not base it on information we have already learned? We rely on the accessibility and correctness of our memories every waking moment; we are frustrated when memory fails (i.e., when we forget) and upset when we learn that something we have long remembered is, in fact, not true.

We usually think of our memories as private and personal; we are the only ones with access to them and they represent the truth about our own lives. But what we remember, and whether we remember it accurately, has both causes and consequences in the real world. The four articles in this section each demonstrate how important memory phenomena from the real world can be studied in the laboratory—and how those laboratory studies then give us insight into how to structure situations in the real world to enhance memory accuracy, memory reliability, and creativity. These articles move between the laboratory, the offices of clinical psychologists, the courtroom, and the business conference room.

In "Imagination and Memory", Garry and Polaschek tackle a hot topic in current memory research—how false memories might arise. There are several types of laboratory studies in which subjects are likely to generate false memories; in many of those studies, subjects are asked to try to remember, visualize, or imagine some event that had not actually occurred in their past. The act of imagining an unreal event can increase people's confidence that the event actually happened.

Erroneous memories can have serious real-world consequences both for the rememberer and for others. In "Recovering Memories of Trauma: A View from the Laboratory", McNally describes attempts to discover the mechanisms behind recovered memories—that is, for memories of traumatic childhood events that have been inaccessible for years and then re-appear in adulthood. Studying recovered memories poses many methodological challenges. McNally describes some differences in personality traits and cognitive processing between people with recovered memories and those with continuous or no memories of childhood trauma. Inaccurate memories can also create problems because the legal system relies extensively on people's memories in criminal proceedings. In "The Confidence of Eyewitnesses in Their Identifications from Lineups", Well, Olson, and Charman describe how the accuracy and confidence of an eyewitness to a crime need not be correlated. An unknown, and probably very large, number of people have been mistakenly convicted of crimes based on faulty eyewitness testimony. The results of psychological research on eyewitness identification has finally begun to have some effect on legal policy.

That memory can be conceptualized as an associative network—a mental web of information constructed such that related concepts are likely to activate each other—may help us understand some of its failings and limitations, including the creation of false memories. However, that conceptualization can also give us ideas about how to overcome blocks in memory and creativity. In "Making Group Brainstorming More Effective: Recommendations from and Associative Memory Perspective," Brown and Paulus use the associative network idea to demonstrate and explain why and how different types of interactions can help to overcome blocking in both individuals and group brainstorming.

Imagination and Memory

Maryanne Garry[1] and Devon L.L. Polaschek
School of Psychology, Victoria University of Wellington, Wellington, New Zealand

Abstract

A growing body of literature shows that imagining contrary-to-truth experiences can change memory. Recent experiments are reviewed to show that when people think about or imagine a false event, entire false memories can be implanted. Imagination inflation can occur even when there is no overt social pressure, and when hypothetical events are imagined only briefly. Overall, studies of imagination inflation show that imagining a counterfactual event can make subjectsmore confident that it actually occurred. We discuss possible mechanisms for imagination inflation and find that, with evidence supporting the involvement of both source confusion and familiarity in creating inflation, the primary mechanism is still to be determined. We briefly review evidence on individual differences in susceptibility to inflation. Finally, the widespread use of imagination-based techniques in self-help and clinical contexts suggests that there may be practical implications when imagination is used as a therapeutic tool.

Keywords

imagination; autobiographical memory

Consistency, said Oscar Wilde, is the last resort of the unimaginative. A poet, playwright, and all-round observer of human behavior, perhaps Wilde knew what psychologists are just beginning to understand: that imagining the past differently from what it was can change the way one remembers it. Psychologists have known, for 15 years or so, that thinking about a different past can change the way people make sense of it. For instance, a person who gets into a car accident may think, "If I hadn't taken this shortcut, I'd never be in this mess now." Perhaps a more common experience for academics is to have a manuscript rejected and think, "If only the editors understood my genius, they would accept this paper."[2] These contrary-to-truth "what if" scenarios, called counterfactual thoughts, can affect the way people judge bad outcomes, personal choices, and accountability (see Roese, 1997, for a review). But an emerging body of literature shows that imagining contrary-to-truth experiences can do more than change the way people make sense of the past. It can also change memories for it.

Loftus (1993) described the first systematic attempt to create in subjects a coherent, detailed memory for things they never did. She simply asked subjects to read detailed descriptions of four events that supposedly happened during their childhood. In actuality, three of these events had been genuinely experienced, and one—about getting lost in a shopping mall—was false. Subjects were asked to write what they remembered about each of these events in separate sessions over the course of a few weeks, and by the last session of the experiment, about 25% of subjects developed a false shopping-mall memory.

It is easy to underestimate the importance of the lost-in-the-mall studies Loftus reported, but even as recently as the early 1990s, there was very little scientific evidence that entire events could systematically be implanted in memory. It was common for nonscientists to charge that laboratory research was not applicable to real events, particularly traumatic ones. It was also common for psychological scientists to argue that it would be difficult to be misled about personally experienced events, especially childhood events.

How did simply reading about being lost in a shopping mall create memories for that event? Loftus (1993) raised a number of possibilities, but it is likely that imagination played a role. Many subjects probably relied on their imagination as a strategy for remembering being lost, because that is what people do when they try to think about an event that they do not remember (Sarbin, 1998).

Hyman and Pentland (1996) used a similar procedure, and compared subjects' performance when they were given a vague instruction to "think about the event" with their performance when they were given an explicit strategic instruction to imagine it. In this research, the unusual false childhood event was about misbehaving at a wedding, knocking over the punch bowl, and spilling punch all over the parents of the bride. One quarter of the "imagine it" subjects and 9% of the "think about it" subjects developed a clear memory of the false experience. Not only do these results demonstrate the effect of imagination in creating memories, but they also suggest that even some "think about it" subjects probably used imagination as a recall strategy.

These studies show that false memories can be created when people think about (and probably imagine) childhood events in an attempt to remember them. However, these studies also relied on social pressure. For example, subjects were told that a family member provided the to-be-remembered events. In some cases, subjects were also interviewed by an experimenter. What if these demands were reduced? Interestingly, research shows that even without them, imagination can still affect memory.

IMAGINATION INFLATION

In a recent study using a less intensive procedure to examine the effect of imagination on memory (Garry, Manning, Loftus, & Sherman, 1996), subjects were pretested on how confident they were that a number of childhood events had happened, asked to imagine some of those events briefly, and then tested again on their confidence that the events had happened. Subjects became more confident they had experienced imagined counterfactual events than nonimagined counterfactual events. This confidence-boosting effect is known as *imagination inflation*. Since this first study, research has consistently shown that briefly imagining the sketchiest details of a counterfactual event is enough to produce imagination inflation (Garry, Frame, & Loftus, 1999; Goff & Roediger, 1998; Heaps & Nash, 1999; Paddock et al., 1998).

What is it about imagining a counterfactual event that causes people to later become more confident that it really happened? The most obvious explanation is that imagining a childhood event does not really do anything interesting at all; it merely reminds people of genuinely experienced events. Indeed, there is no way to tell how much of the imagination inflation is caused by simple reminding in the studies in which social pressures to remember are reduced (Garry et al., 1996). However, Goff and Roediger (1998) addressed the issue by asking subjects first to do some actions but not others. Later, subjects imagined some of those actions anywhere from zero to five times. Goff and Roediger found evidence of imagination inflation, as well as an effect for the number of imaginings.

There are two main explanations for imagination inflation. The first of these is source confusion (Johnson, Hashtroudi, & Lindsay, 1993), a process in which content and source (i.e., the circumstances in which information was learned) become separated. In a source-confusion account of imagination inflation, subjects are said to confuse information from a recently imagined event with information from a genuinely experienced event. The other explanation for imagination inflation is familiarity. In this account of imagination inflation, imagining events makes them more familiar, and subjects incorrectly attribute the increased familiarity of imagined events to their occurrence (see also Jacoby, Kelley, & Dywan, 1989).

Goff and Roediger (1998) argued that imagination inflation is the product of overlapping source confusion and familiarity mechanisms, but a recent experiment (Garry et al., 1999) tried to separate the two. Subjects were asked to imagine either certain childhood events happening to them or those same childhood events happening to another person. Subjects in the two groups showed the same amount of imagination inflation. Moreover, subjects' rating of various imagery qualities and characteristics did not predict the amount of imagination inflation. Both of these findings are evidence that imagination inflation is caused more by familiarity than by source confusion.

Of course, there are alternative explanations for these results. One is that subjects may have pretended to be the other character; for example, if Beth pretended to be an 8-year-old Bill Clinton breaking a window with his hand, the images she has of the event would be from young Bill's perspective, and thus as easily confused with a real event as if Beth imagined breaking the window as a child herself. So the results do not necessarily indicate that source confusion plays no role in imagination inflation. Indeed, evidence for the primary role of source confusion in imagination inflation is found in studies that have varied the time distance of hypothetical events. A source-confusion mechanism predicts greater imagination inflation for long-ago imagined events compared with more recent imagined events, whereas a familiarity mechanism predicts no difference in the amount of imagination inflation. These opposing predictions led to experiments (Garry & Hayes, 1999) in which some subjects imagined hypothetical childhood events and others imagined more recent events from only 5 years ago. The subjects who imagined the long-ago childhood events showed the typical imagination-inflation effect, but those who were asked to imagine the recent events showed no change in confidence.

WHO IS SUSCEPTIBLE TO IMAGINATION INFLATION?

Are some people more likely to inflate than others? This is a question several researchers have asked. Heaps and Nash (1999) found that subjects' tendency toward imagination inflation had nothing to do with their susceptibility to the influence of an authoritative questioner, nor with the vividness of their mental images. Instead, it was their predisposition to both hypnotic suggestion and dissociation (i.e., tendency to lose awareness and confuse fact with fantasy) that predicted imagination inflation, a finding supported by the work of Paddock et al. (1998). Paddock et al. also suggested that age may be inversely related to imagination inflation. It is not certain if age and education somehow modulate the personal characteristics that predispose people to imagination inflation; much research suggests, for instance, that both hypnotic suggestibility and dissociativity decrease with age.

Of course, age is surely the most contentious of the individual differences considered in the false memory literature. Researchers studying the suggestibility of children have long been aware of the dangerous consequences of thinking about a false event. Much research shows that children can adopt misleading suggestions about aspects of genuine events, or come to report entire, elaborate false events (Ceci, Loftus, Leichtman, & Bruck, 1994). Are children susceptible to imagination inflation? When 8- to 10-year-olds imagined hypothetical events from 5 years earlier, the children showed inflated confidence that the events had really happened (Garry & Hayes, 1999).

THEORETICAL IMPLICATIONS

A growing literature shows that imagination can change autobiographies. The most important theoretical implication of this research is the Heisenberg-like suggestion that repeatedly examining one's past can affect the way one remembers it. In the imagination-inflation literature, several important questions remain. It is still not clear, for instance, what the primary mechanism driving imagination inflation is, nor is it known under what circumstances imagining recent autobiographical events might boost confidence that they occurred. Interestingly, Goff and Roediger (1998) actually found no imagination effect when recent events were imagined only once, the condition that most closely parallels the standard imagination-inflation procedure, which examines naturalistic, autobiographical events.

Finally, there are two statistical issues to consider in making sense of imagination inflation. First, although in one sense it is impressive that a seemingly innocuous, brief imagination can affect people's confidence in whether or not an event actually occurred, from a statistical point of view, the imagination-inflation effect is a small one. Future research should look toward increasing the size of the imagination-inflation effect; there are aspects of the current experimental procedures that might be masking or moderating the effect. Second, one might at first glance be tempted to argue that the effect is nothing other than "regression toward the mean," a statistical phenomenon in which people with extremely low scores on some measure (in this case, confidence in whether an event actually occurred) will move toward the population mean on that measure when tested again. In the original imagination-inflation research (Garry et al., 1996), we selected subjects with low confidence scores, and on a retest their scores were higher. However, an interpretation of the results in terms of regression to the mean is easily dismissed, because the same people participated in both the imagination condition and the control, no-imagination condition, and confidence did not rise as much in the control condition as in the imagination condition. Moreover, in both conditions, pretest scores were equally low on the critical events. Thus, regression toward the mean cannot explain the imagination-inflation effect.

CLINICAL IMPLICATIONS

There are practical implications of this research, too. Although the clinical implications of the lost-in-the-mall genre of studies are obvious, what may be less obvious is that the results of the imagination studies (e.g., Garry et al., 1996) also give cause for concern. The imagination-inflation literature may have implications for therapists who are involved in a much broader range of clinical endeavors than working with survivors of sexual abuse. In particular, Goff and Roediger's (1998) findings show that even when people imagine performing certain actions in the present, they can report, mistakenly, that they genuinely did them.

A review of the psychotherapy and self-help literatures suggests that imagined performance features in a variety of therapeutic strategies. For example, cognitive behavioral psychologists use a number of techniques that rely heavily on the use of imagination, particularly in the treatment of anxiety disorders. If Bo Peep is instructed to imagine a series of sheep-related feared situations to reduce her sheep phobia, she will probably become less anxious about sheep, but she may also show inflated confidence that she actually experienced sheep encounters that in fact she only imagined.

The self-help literature is also replete with imagination-based interventions. Books on creative visualization exhort readers to imagine themselves achieving anything from younger-looking skin to becoming president of the United States. In more academic circles, researchers have found that mental simulation can enhance coping with stressful events and improve athletic performance (Orlick & Partington, 1986; Taylor, Pham, Rivkin, & Armor, 1998). These techniques rely on repeatedly imagining situations and actions.

Of course, the unanticipated memory-related consequences suggested by imagination-inflation research may not always be serious. Incorrectly recalling that a particular experience or action actually occurred although it was only imagined may not matter outside of evidential contexts. Nevertheless, good clinical practice requires that clinicians understand both how efficacious treatments work and what their possible side effects are, and further research should be aimed at informing both science and practice. Meanwhile, the growing literature on imagination and memory suggests that there can be some unexpected side effects to using imagination as a therapeutic tool. Inconsistency, at least in the way people remember their own lives, may be the price people pay for being imaginative.

Recommended Reading

Garry, M., Frame, S., & Loftus, E.F. (1999). (See References)
Loftus, E.F. (1997, September). Creating false memories. *Scientific American*, 277, 70–75.

Ofshe, R., & Watters, E. (1994). *Making monsters: False memories, psychotherapy, and sexual hysteria.* Berkeley: University of California.

Notes

1. Address correspondence to Maryanne Garry, School of Psychology, Victoria University of Wellington, Box 600, Wellington, New Zealand.

2. Obviously, we are speaking here about our own experiences.

References

Ceci, S.J., Loftus, E.F., Leichtman, M.D., & Bruck, M. (1994). The possible role of source misattributions in the creation of false beliefs among preschoolers. *International Journal of Clinical and Experimental Hypnosis, 42*, 304–320.

Garry, M., Frame, S., & Loftus, E.F. (1999). Lie down and let me tell you about your childhood. In S. Della Sala (Ed.), *Mind myths: Exploring popular assumptions about the mind and brain* (pp. 113–124). New York: Wiley.

Garry, M., & Hayes, J.H. (1999). *Imagination inflation depends on when the imagined event occurred.* Unpublished manuscript, Victoria University of Wellington, Wellington, New Zealand.

Garry, M., Manning, C.G., Loftus, E.F., & Sherman, S.J. (1996). Imagination inflation: Imagining a childhood event inflates confidence that it occurred. *Psychonomic Bulletin & Review, 3*, 208–214.

Goff, L.M., & Roediger, H.L., III. (1998). Imagination inflation for action events: Repeated imaginings lead to illusory recollections. *Memory & Cognition, 26*, 20–33.

Heaps, C., & Nash, M.R. (1999). Individual differences in imagination inflation. *Psychonomic Bulletin & Review, 6*, 313–318.

Hyman, I.E., & Pentland, J. (1996). The role of mental imagery in the creation of false childhood memories. *Journal of Memory and Language, 35*, 101–117.

Jacoby, L.L., Kelley, C.M., & Dywan, J. (1989). Memory attributions. In H.L. Roediger, III, & F.I.M. Craik (Eds.), *Varieties of memory and consciousness: Essays in honour of Endel Tulving* (pp. 391–422). Hillsdale, NJ: Erlbaum.

Johnson, M.K., Hashtroudi, S., & Lindsay, D.S. (1993). Source monitoring. *Psychological Bulletin, 114*, 3–28.

Loftus, E.F. (1993). The reality of repressed memories. *American Psychologist, 48*, 518–537.

Orlick, T., & Partington, J.T. (1986). *Psyched: Inner views of winning.* Ottawa, Canada: Coaching Association of Canada.

Paddock, J.R., Joseph, A.L., Chan, F.M., Terranova, S., Loftus, E.F., & Manning, C. (1998). When guided visualization procedures may backfire: Imagination inflation and predicting individual differences in suggestibility. *Applied Cognitive Psychology, 12*, S63–S75.

Roese, N.J. (1997). Counterfactual thinking. *Psychological Bulletin, 121*, 133–148.

Sarbin, T.R. (1998). Believed-in imaginings: A narrative approach. In J. de Rivera & T.R. Sarbin (Eds.), *Believed-in imaginings: The narrative construction of reality* (pp. 15–30). Washington, DC: American Psychological Association.

Taylor, S.E., Pham, L.B., Rivkin, I.D., & Armor, D.A. (1998). Harnessing the imagination: Mental simulation, self-regulation, and coping. *American Psychologist, 53*, 429–439.

Critical Thinking Questions

1. Think back to your 4th (or 3rd or 5th) birthday. Close your eyes and try to imagine it. Was there a party? Was there a cake with candles for you to blow out? Presents? (a) Can you envision that thinking about your birthday this way might cause you to believe that you had a birthday party when you really hadn't? Why or why not? (b) Perhaps you do have memories of an early birthday. Are you sure those memories are your own? Might those memories come

from stories that older people (e.g., your parents or siblings) have told you? Might they come from having seen pictures or videos of that birthday? How can you be sure?

2. One of the editors of this book has imagined herself being Scarlett O'Hara (from Gone With the Wind) dozens of times. Yet she does not believe that she has ever owned slaves, picked cotton, or worn a dress made of her mother's curtains. The authors argue that familiarity may be the mechanism behind false memories and imagination inflation, but is familiarity sufficient? In Question 1 above, do you think you might ever believe that: You got a (real) giraffe as a gift? Two Martians came to your party? What do you think the limits are to the kinds of events that for which memory could be affected by imagination? What might that suggest about people's recovered memories for things like satanic rituals?
3. Imagination inflation has been shown to operate more for long-ago events than for recent events. Why might that happen?

Recovering Memories of Trauma: A View From the Laboratory

Richard J. McNally[1]

Department of Psychology, Harvard University, Cambridge, Massachusetts

Abstract

The controversy over the validity of repressed and recovered memories of childhood sexual abuse (CSA) has been extraordinarily bitter. Yet data on cognitive functioning in people reporting repressed and recovered memories of trauma have been strikingly scarce. Recent laboratory studies have been designed to test hypotheses about cognitive mechanisms that ought to be operative if people can repress and recover memories of trauma or if they can form false memories of trauma. Contrary to clinical lore, these studies have shown that people reporting CSA histories are not characterized by a superior ability to forget trauma-related material. Other studies have shown that individuals reporting recovered memories of either CSA or abduction by space aliens are characterized by heightened proneness to form false memories in certain laboratory tasks. Although cognitive psychology methods cannot distinguish true memories from false ones, these methods can illuminate mechanisms for remembering and forgetting among people reporting histories of trauma.

Keywords

recovered memories; trauma; repression; sexual abuse; dissociation

How victims remember trauma is among the most explosive issues facing psychology today. Most experts agree that combat, rape, and other horrific experiences are unforgettably engraved on the mind (Pope, Oliva, & Hudson, 1999). But some also believe that the mind can defend itself by banishing traumatic memories from awareness, making it difficult for victims to remember them until many years later (Brown, Scheflin, & Hammond, 1998).

This controversy has spilled out of the clinics and cognitive psychology laboratories, fracturing families, triggering legislative change, and determining outcomes in civil suits and criminal trials. Most contentious has been the claim that victims of childhood sexual abuse (CSA) often repress and then recover memories of their trauma in adulthood.[2] Some psychologists believe that at least some of these memories may be false—inadvertently created by risky therapeutic methods (e.g., hypnosis, guided imagery; Ceci & Loftus, 1994).

One striking aspect of this controversy has been the paucity of data on cognitive functioning in people reporting repressed and recovered memories of CSA. Accordingly, my colleagues and I have been conducting studies designed to test hypotheses about mechanisms that might enable people either to repress and recover memories of trauma or to develop false memories of trauma.

For several of our studies, we recruited four groups of women from the community. Subjects in the *repressed-memory group* suspected they had been sexually abused as children, but they had no explicit memories of abuse. Rather, they inferred their hidden abuse history from diverse indicators, such as depressed mood, interpersonal problems with men, dreams, and brief, recurrent visual images (e.g., of a penis), which they interpreted as "flashbacks" of early trauma. Subjects in the *recovered-memory group* reported having remembered their abuse after long periods of not having thought about it.[3] Unable to corroborate their reports, we cannot say whether the memories were true or false. Lack of corroboration, of course, does not mean that a memory is false. Subjects in the *continuous-memory group* said that they had never forgotten their abuse, and subjects in the *control group* reported never having been sexually abused.

PERSONALITY TRAITS AND PSYCHIATRIC SYMPTOMS

To characterize our subjects in terms of personality traits and psychiatric symptoms, we asked them to complete a battery of questionnaires measuring normal personality variation (e.g., differences in absorption, which includes the tendency to fantasize and to become emotionally engaged in movies and literature), depressive symptoms, posttraumatic stress disorder (PTSD) symptoms, and dissociative symptoms (alterations in consciousness, such as memory lapses, feeling disconnected with one's body, or episodes of "spacing out"; McNally, Clancy, Schacter, & Pitman, 2000b).

There were striking similarities and differences among the groups in terms of personality profiles and psychiatric symptoms. Subjects who had always remembered their abuse were indistinguishable from those who said they had never been abused on all personality measures. Moreover, the continuous-memory and control groups did not differ in their symptoms of depression, posttraumatic stress, or dissociation. However, on the measure of negative affectivity—proneness to experience sadness, anxiety, anger, and guilt—the repressed-memory group scored higher than did either the continuous-memory or the control group, whereas the recovered-memory group scored midway between the repressed-memory group on the one hand and the continuous-memory and control groups on the other.

The repressed-memory subjects reported more depressive, dissociative, and PTSD symptoms than did continuous-memory and control subjects. Repressed-memory subjects also reported more depressive and PTSD symptoms than did recovered-memory subjects, who, in turn, reported more dissociative and PTSD symptoms than did control subjects. Finally, the repressed and recovered-memory groups scored higher than the control group on the measure of fantasy proneness, and the repressed-memory group scored higher than the continuous-memory group on this measure.

This psychometric study shows that people who believe they harbor repressed memories of sexual abuse are more psychologically distressed than those who say they have never forgotten their abuse.

FORGETTING TRAUMA RELATED MATERIAL

Some clinical theorists believe that sexually molested children learn to disengage their attention during episodes of abuse and allocate it elsewhere (e.g., Terr,

1991). If CSA survivors possess a heightened ability to disengage attention from threatening cues, impairing their subsequent memory for them, then this ability ought to be evident in the laboratory. In our first experiment, we used directed-forgetting methods to test this hypothesis (McNally, Metzger, Lasko, Clancy, & Pitman, 1998). Our subjects were three groups of adult females: CSA survivors with PTSD, psychiatrically healthy CSA survivors, and nonabused control subjects. Each subject was shown, on a computer screen, a series of words that were either trauma related (e.g., *molested*), positive (e.g., *charming*), or neutral (e.g., *mailbox*). Immediately after each word was presented, the subject received instructions telling her either to remember the word or to forget it. After this encoding phase, she was asked to write down all the words she could remember, irrespective of the original instructions that followed each word.

If CSA survivors, especially those with PTSD, are characterized by heightened ability to disengage attention from threat cues, thereby attenuating memory for them, then the CSA survivors with PTSD in this experiment should have recalled few trauma words, especially those they had been told to forget. Contrary to this hypothesis, this group exhibited memory deficits for positive and neutral words they had been told to remember, while demonstrating excellent memory for trauma words, including those they had been told to forget. Healthy CSA survivors and control subjects recalled remember-words more often than forget-words regardless of the type of word. Rather than possessing a superior ability to forget trauma-related material, the most distressed survivors exhibited difficulty banishing this material from awareness.

In our next experiment, we used this directed-forgetting approach to test whether repressed- and recovered-memory subjects, relative to nonabused control subjects, would exhibit the hypothesized superior ability to forget material related to trauma (McNally, Clancy, & Schacter, 2001). If anyone possesses this ability, it ought to be such individuals. However, the memory performance of the repressed- and recovered-memory groups was entirely normal: They recalled remember-words better than forget-words, regardless of whether the words were positive, neutral, or trauma related.

INTRUSION OF TRAUMATIC MATERIAL

The hallmark of PTSD is involuntary, intrusive recollection of traumatic experiences. Clinicians have typically relied on introspective self-reports as confirming the presence of this symptom. The emotional Stroop color-naming task provides a quantitative, nonintrospective measure of intrusive cognition. In this paradigm, subjects are shown words varying in emotional significance, and are asked to name the colors the words are printed in while ignoring the meanings of the words. When the meaning of a word intrusively captures the subject's attention despite the subject's efforts to attend to its color, Stroop interference—delay in color naming—occurs. Trauma survivors with PTSD take longer to name the colors of words related to trauma than do survivors without the disorder, and also take longer to name the colors of trauma words than to name the colors of positive and neutral words or negative words unrelated to their trauma (for a review, see McNally, 1998).

Using the emotional Stroop task, we tested whether subjects reporting either continuous, repressed, or recovered memories of CSA would exhibit interference for trauma words, relative to nonabused control subjects (McNally, Clancy, Schacter, & Pitman, 2000a). If severity of trauma motivates repression of traumatic memories, then subjects who cannot recall their presumably repressed memories may nevertheless exhibit interference for trauma words. We presented a series of trauma-related, positive, and neutral words on a computer screen, and subjects named the colors of the words as quickly as possible. Unlike patients with PTSD, including children with documented abuse histories (Dubner & Motta, 1999), none of the groups exhibited delayed color naming of trauma words relative to neutral or positive ones.

MEMORY DISTORTION AND FALSE MEMORIES IN THE LABORATORY

Some psychotherapists who believe their patients suffer from repressed memories of abuse will ask them to visualize hypothetical abuse scenarios, hoping that this guided-imagery technique will unblock the presumably repressed memories. Unfortunately, this procedure may foster false memories.

Using Garry, Manning, Loftus, and Sherman's (1996) methods, we tested whether subjects who have recovered memories of abuse are more susceptible than control subjects to this kind of memory distortion (Clancy, McNally, & Schacter, 1999). During an early visit to the laboratory, subjects rated their confidence regarding whether they had experienced a series of unusual, but nontraumatic, childhood events (e.g., getting stuck in a tree). During a later visit, they performed a guided-imagery task requiring them to visualize certain of these events, but not others. They later rerated their confidence that they had experienced each of the childhood events. Nonsignificant trends revealed an inflation in confidence for imagined versus non-imagined events. But the magnitude of this memory distortion was more than twice as large in the control group as in the recovered memory group, contrary to the hypothesis that people who have recovered memories of CSA would be especially vulnerable to the memory-distorting effects of guided imagery.

To use a less-transparent paradigm for assessing proneness to develop false memories, we adapted the procedure of Roediger and McDermott (1995). During the encoding phase in this paradigm, subjects hear word lists, each consisting of semantically related items (e.g., *sour*, *bitter*, *candy*, sugar) that converge on a nonpresented word—the *false target*—that captures the gist of the list (e.g., *sweet*). On a subsequent recognition test, subjects are given a list of words and asked to indicate which ones they heard during the previous phase. The false memory effect occurs when subjects "remember" having heard the false target. We found that recovered-memory subjects exhibited greater proneness to this false memory effect than did subjects reporting either repressed memories of CSA, continuous memories of CSA, or no abuse (Clancy, Schacter, McNally, & Pitman, 2000). None of the lists was trauma related, and so we cannot say whether the effect would have been more or less pronounced for words directly related to sexual abuse.

In our next experiment, we tested people whose memories were probably false: individuals reporting having been abducted by space aliens (Clancy, McNally, Schacter, Lenzenweger, & Pitman, 2002). In addition to testing these individuals (and control subjects who denied having been abducted by aliens), we tested individuals who believed they had been abducted, but who had no memories of encountering aliens. Like the repressed-memory subjects in our previous studies, they inferred their histories of trauma from various "indicators" (e.g., a passion for reading science fiction, unexplained marks on their bodies). Like subjects with recovered memories of CSA, those reporting recovered memories of alien abduction exhibited pronounced false memory effects in the laboratory. Subjects who only believed they had been abducted likewise exhibited robust false memory effects.

CONCLUSIONS

The aforementioned experiments illustrate one way of approaching the recovered-memory controversy. Cognitive psychology methods cannot ascertain whether the memories reported by our subjects were true or false, but these methods can enable testing of hypotheses about mechanisms that ought to be operative if people can repress and recover memories of trauma or if they can develop false memories of trauma.

Pressing issues remain unresolved. For example, experimentalists assume that directed forgetting and other laboratory methods engage the same cognitive mechanisms that generate the signs and symptoms of emotional disorder in the real world. Some therapists question the validity of this assumption. Surely, they claim, remembering or forgetting the word *incest* in a laboratory task fails to capture the sensory and narrative complexity of autobiographical memories of abuse. On the one hand, the differences between remembering the word *incest* in a directed-forgetting experiment, for example, and recollecting an episode of molestation do, indeed, seem to outweigh the similarities. On the other hand, laboratory studies may underestimate clinical relevance. For example, if someone cannot expel the word *incest* from awareness during a directed-forgetting experiment, then it seems unlikely that this person would be able to banish autobiographical memories of trauma from consciousness. This intuition notwithstanding, an important empirical issue concerns whether these tasks do, indeed, engage the same mechanisms that figure in the cognitive processing of traumatic memories outside the laboratory.

A second issue concerns attempts to distinguish subjects with genuine memories of abuse from those with false memories of abuse. Our group is currently exploring whether this might be done by classifying trauma narratives in terms of how subjects describe their memory-recovery experience. For example, some of the subjects in our current research describe their recovered memories of abuse by saying, "I had forgotten about that. I hadn't thought about the abuse in years until I was reminded of it recently." The narratives of other recovered-memory subjects differ in their experiential quality. These subjects, as they describe it, suddenly realize that they are abuse survivors, sometimes attributing current life difficulties to these long-repressed memories. That is, they do not say that they have remembered forgotten events they once knew, but rather indicate that they have learned (e.g., through hypnosis) the abuse occurred. It will be important to determine whether these two groups of recovered-memory subjects differ cognitively. For example, are subjects exemplifying the second type of recovered-memory experience more prone to develop false memories in the laboratory than are subjects exemplifying the first type of experience?

Recommended Reading

Lindsay, D.S., & Read, J.D. (1994). Psychotherapy and memories of childhood sexual abuse: A cognitive perspective. *Applied Cognitive Psychology*, *8*, 281–338.

McNally, R.J. (2001). The cognitive psychology of repressed and recovered memories of childhood sexual abuse: Clinical implications. *Psychiatric Annals*, *31*, 509–514.

McNally, R.J. (2003). Progress and controversy in the study of posttraumatic stress disorder. *Annual Review of Psychology*, *54*, 229–252.

McNally, R.J. (2003). *Remembering trauma.* Cambridge, MA: Belknap Press/Harvard University Press.

Piper, A., Jr., Pope, H.G., Jr., & Borowiecki, J.J., III. (2000). Custer's last stand: Brown, Scheflin, and Whitfield's latest attempt to salvage "dissociative amnesia." *Journal of Psychiatry and Law*, *28*, 149–213.

Acknowledgments—Preparation of this article was supported in part by National Institute of Mental Health Grant MH61268.

Notes

1. Address correspondence to Richard J. McNally, Department of Psychology, Harvard University, 1230 William James Hall, 33 Kirkland St., Cambridge, MA 02138; e-mail: rjm@wjh. harvard.edu.

2. Some authors prefer the term *dissociation* (or *dissociative amnesia*) to *repression*. Although these terms signify different proposed mechanisms, for practical purposes these variations make little difference in the recovered-memory debate. Each term implies a defensive process that blocks access to disturbing memories.

3. However, not thinking about a disturbing experience for a long period of time must not be equated with an inability to remember it. Amnesia denotes an inability to recall information that has been encoded.

References

Brown, D., Scheflin, A.W., & Hammond, D.C. (1998). *Memory, trauma treatment, and the law.* New York: Norton.

Ceci, S.J., & Loftus, E.F. (1994). 'Memory work': A royal road to false memories? *Applied Cognitive Psychology*, *8*, 351–364.

Clancy, S.A., McNally, R.J., & Schacter, D.L. (1999). Effects of guided imagery on memory distortion in women reporting recovered memories of childhood sexual abuse. *Journal of Traumatic Stress*, *12*, 559–569.

Clancy, S.A., McNally, R.J., Schacter, D.L., Lenzenweger, M.F., & Pitman, R.K. (2002). Memory distortion in people reporting abduction by aliens. *Journal of Abnormal Psychology*, *111*, 455–461.

Clancy, S.A., Schacter, D.L., McNally, R.J., & Pitman, R.K. (2000). False recognition in women reporting recovered memories of sexual abuse. *Psychological Science*, *11*, 26–31.

Dubner, A.E., & Motta, R.W. (1999). Sexually and physically abused foster care children and posttraumatic stress disorder. *Journal of Consulting and Clinical Psychology*, *67*, 367–373.

Garry, M., Manning, C.G., Loftus, E.F., & Sherman, S.J. (1996). Imagination inflation: Imagining a childhood event inflates confidence that it occurred. *Psychonomic Bulletin & Review, 3*, 208–214.

McNally, R.J. (1998). Experimental approaches to cognitive abnormality in posttraumatic stress disorder. *Clinical Psychology Review, 18*, 971–982.

McNally, R.J., Clancy, S.A., & Schacter, D.L. (2001). Directed forgetting of trauma cues in adults reporting repressed or recovered memories of childhood sexual abuse. *Journal of Abnormal Psychology, 110*, 151–156.

McNally, R.J., Clancy, S.A., Schacter, D.L., & Pitman, R.K. (2000a). Cognitive processing of trauma cues in adults reporting repressed, recovered, or continuous memories of childhood sexual abuse. *Journal of Abnormal Psychology, 109*, 355–359.

McNally, R.J., Clancy, S.A., Schacter, D.L., & Pitman, R.K. (2000b). Personality profiles, dissociation, and absorption in women reporting repressed, recovered, or continuous memories of childhood sexual abuse. *Journal of Consulting and Clinical Psychology, 68*, 1033–1037.

McNally, R.J., Metzger, L.J., Lasko, N.B., Clancy, S.A., & Pitman, R.K. (1998). Directed forgetting of trauma cues in adult survivors of childhood sexual abuse with and without posttraumatic stress disorder. *Journal of Abnormal Psychology, 107*, 596–601.

Pope, H.G., Jr., Oliva, P.S., & Hudson, J.I. (1999). Repressed memories: The scientific status. In D.L. Faigman, D.H. Kaye, M.J. Saks, & J. Sanders (Eds.), *Modern scientific evidence: The law and science of expert testimony* (Vol. 1, pocket part, pp. 115–155). St. Paul, MN: West Publishing.

Roediger, H.L., III, & McDermott, K.B. (1995). Creating false memories: Remembering words not presented in lists. *Journal of Experimental Psychology: Learning, Memory, and Cognition, 21*, 803–814.

Terr, L.C. (1991). Childhood traumas: An outline and overview. *American Journal of Psychiatry, 148*, 10–20.

Critical Thinking Questions

1. Studying recovered memory provides many challenges. A major difficulty, of course, is that the researcher is almost never certain that the remembered event(s) actually occurred. In some of the present research, the groups studied were recovered-memory, repressed-memory, continuous-memory, and control. Why is it important to have each of those groups? Why was it important to do similar studies with the CSA+PTSD, CSA, and control groups? And, silly as it may seem, why was it important to do similar studies on people with memories of alien abduction?

2. Another difficulty is provided by ethical constraints in research. Psychologists can't randomly select a bunch of people to traumatize and then assign them to the recovered-memory condition. What implications does that limitation have for the causal direction of the finding that the recovered-memory and repressed-memory groups differ on some personality dimensions from the continuous-memory and control groups?

3. The article describes four different experimental procedures – directed forgetting, emotional Stroop, imagination inflation, and the false-target list. Only on the last one did the recovered-memory group significantly differ from the others. To what extent do you believe that each of these procedures captures something about the mechanisms behind memory repression? What about the mechanisms behind how false memories might occur for real-life traumatic events that occurred over a long period of time?

4. Suppose you read in the newspaper about three women who claimed to have recently recovered memories for childhood sexual abuse. One remembered the abuse after months of therapy for relationship issues in which the therapist suggested that she try to imagine how she might have been sexually abused as a child. The second remembered the abuse when she realized it made sense as an explanation for a pattern of beliefs and behaviors in her life (e.g., why she was uncomfortable in sexual relationships; why she had body-image problems). The third remembered the abuse out of the blue when she entered a child's bedroom that had the same wallpaper as had been in her childhood bedroom. Of course, all three memories could be true, but which are you most and least likely to believe. Why?

The Confidence of Eyewitnesses in Their Identifications From Lineups

Gary L. Wells,[1] Elizabeth A. Olson,
and Steve D. Charman
Psychology Department, Iowa State University, Ames, Iowa

Abstract

The confidence that eyewitnesses express in their lineup identifications of criminal suspects has a large impact on criminal proceedings. Many convictions of innocent people can be attributed in large part to confident but mistaken eyewitnesses. Although reasonable correlations between confidence and accuracy can be obtained under certain conditions, confidence is governed by some factors that are unrelated to accuracy. An understanding of these confidence factors helps establish the conditions under which confidence and accuracy are related and leads to important practical recommendations for criminal justice proceedings.

Keywords

eyewitness testimony; lineups; eyewitness memory

Mistaken identification by eyewitnesses was the primary evidence used to convict innocent people whose convictions were later overturned by forensic DNA tests (Scheck, Neufeld, & Dwyer, 2000; Wells et al., 1998). The eyewitnesses in these cases were very persuasive because on the witness stand they expressed extremely high confidence that they had identified the actual perpetrator. Long before DNA exoneration cases began unfolding in the 1990s, however, eyewitness researchers in psychology were finding that confidence is not a reliable indicator of accuracy and warning the justice system that heavy reliance on eyewitness's confidence in their identifications might lead to the conviction of innocent people.

Studies have consistently demonstrated that the confidence an eyewitness expresses in an identification is the major factor determining whether people will believe that the eyewitness made an accurate identification. The confidence an eyewitness expresses is also enshrined in the criteria that the U.S. Supreme Court used 30 years ago (and that now guide lower courts) for deciding the accuracy of an eyewitness's identification in a landmark case. Traditionally, much of the experimental work examining the relation between confidence and accuracy in eyewitness identification tended to frame the question as "What is the correlation between confidence and accuracy?" as though there were some single, true correlation value. Today, eyewitness researchers regard the confidence-accuracy relation as something that varies across circumstances. Some of these circumstances are outside the control of the criminal justice system, but some are determined by the procedures that criminal justice personnel control.

A GENERAL FRAMEWORK FOR CONFIDENCE ACCURACY RELATIONS

It has been fruitful to think about eyewitness accuracy and eyewitness confidence as variables that are influenced by numerous factors, some of which are the same and some of which are different. We expect confidence and accuracy to be more closely related when the variables that are influencing accuracy are also influencing confidence than when the variables influencing accuracy are different from those influencing confidence. Consider, for instance, the variable of exposure duration (i.e., how long the eyewitness viewed the culprit while the crime was committed). An eyewitness who viewed the culprit for a long time during the crime should be more accurate than one who had only a brief view. Furthermore, the longer view could be a foundation for the eyewitness to feel more confident in the identification, either because the witness has a more vivid and fluent memory from the longer duration or because the witness infers his or her accuracy from the long exposure duration. Hence, the correlation between confidence and accuracy should be higher the more variation there is in the exposure duration across witnesses (Read, Vokey, & Hammersley, 1990). Suppose, however, that some eyewitnesses were reinforced after their identification decision (e.g., "Good job. You are a good witness."), whereas others were given no such reinforcement. Such postidentification reinforcement does nothing to make witnesses more accurate, but dramatically inflates their confidence (Wells & Bradfield, 1999).

Eyewitness confidence can be construed simply as the eyewitness's belief, which varies in degree, about whether the identification was accurate or not. This belief can have various sources, both internal and external, that need not be related to accuracy. Shaw and his colleagues, for example, have shown that repeated questioning of eyewitnesses about mistaken memories does not make the memories more accurate but does inflate the eyewitnesses' confidence in those memories (Shaw, 1996; Shaw & McLure, 1996). Although the precise mechanisms for the repeated-questioning effect are not clear (e.g., increased commitment to the mistaken memory vs. increased fluidity of the response), these results illustrate a dissociation between variables affecting confidence and variables affecting accuracy.

It is useful to think about broad classes of variables that could be expected to drive confidence and not accuracy, or to drive accuracy and not confidence, or to drive both variables. It is even possible to think about variables that could decrease accuracy while increasing confidence. Consider, for instance, coincidental resemblance. Mistaken identifications from lineups occur primarily when the actual culprit is not in the lineup. Suppose there are two such lineups, one in which the innocent suspect does not highly resemble the real culprit and a second in which the innocent suspect is a near clone (coincidental resemblance) of the real culprit. The second lineup will result not only in an increased rate of mistaken identification compared with the first lineup, but also in higher confidence in that mistake. In this case, a variable that decreases accuracy (resemblance of an innocent suspect to the actual culprit) serves to increase confidence.

THE CORRELATION, CALIBRATION, AND INFLATION OF CONFIDENCE

Although many individual studies have reported little or no relation between eyewitnesses' confidence in their identifications and the accuracy of their identifications, an analysis that statistically combined individual studies indicates that the confidence-accuracy correlation might be as high as +.40 when the analysis is restricted to individuals who make an identification (vs. all witnesses; see Sporer, Penrod, Read, & Cutler, 1995). How useful is this correlation for predicting accuracy from confidence? In some ways, a correlation of .40 could be considered strong. For instance, when overall accuracy is 50%, a .40 correlation would translate into 70% of the witnesses with high confidence being accurate and only 30% of the witnesses with low confidence being accurate. As accuracy deviates from 50%, however, differences in accuracy rates between witnesses with high and low confidence will diminish even though the correlation remains .40.

Another way to think about a .40 correlation is to compare it with something that people experience in daily life, namely the correlation between a person's height and a person's gender. Extrapolating from males' and females' average height and standard deviation (69.1, 63.7, and 5.4 in., respectively; Department of Health and Human Services, n.d.) yields a correlation between height and gender of +.43. Notice that the correlation between height and gender is quite similar to the correlation between eyewitnesses' identification confidence and accuracy. Thus, if eyewitnesses' identifications are accurate 50% of the time, we would expect to encounter a highly confident mistaken eyewitness (or a nonconfident accurate eyewitness) about as often as we would encounter a tall female (or a short male).

Although the eyewitness-identification literature has generally used correlation methods to express the statistical association between confidence and accuracy, it is probably more forensically valid to use calibration and overconfidence/underconfidence measures rather than correlations (Brewer, Keast, & Rishworth, 2002; Juslin, Olson, & Winman, 1996). In effect, the correlation method (specifically, point-biserial correlation) expresses the degree of statistical association by calculating the difference in confidence (expressed in terms of the standard deviation) between accurate and inaccurate witnesses. Calibration, on the other hand, assesses the extent to which an eyewitness's confidence, expressed as a percentage, matches the probability that the eyewitness is correct. Overconfidence reflects the extent to which the percentage confidence exceeds the probability that the eyewitness is correct (e.g., 80% confidence and 60% probability correct), and underconfidence reflects the extent to which the percentage confidence underestimates the probability that the eyewitness is correct (e.g., 40% confidence and 60% probability correct). Juslin et al. pointed out that the confidence-accuracy correlation can be quite low even when calibration is high.

Work by Juslin et al. (1996) indicates that eyewitnesses can be well calibrated at times, but recent experiments (Wells & Bradfield, 1999) illustrate a problem that can arise when trying to use percentage confidence expressed by witnesses to infer the probability that their identifications are accurate. In a series of experiments, eyewitnesses were induced to make mistaken identifications from lineups in which the culprit was absent and were then randomly assigned to receive confirming "feedback" telling them that they identified the actual suspect or to receive no feedback at all. Later, these witnesses were asked how certain they were at the time of their identification (i.e., how certain they were before the feedback). Those who did not receive confirming feedback gave average confidence ratings of less than 50%, but those receiving confirming feedback gave average confidence ratings of over 70%. Because all of these eyewitnesses had made mistaken identifications, even the no-feedback witnesses were overconfident, but the confirming-feedback witnesses were especially overconfident. Confidence inflation is a difficult problem in actual criminal cases because eyewitnesses are commonly given feedback about whether their identification decisions agree with the investigator's theory of the case. In these cases, it is the detective, rather than the eyewitness, who determines the confidence of the eyewitness.

Confirming feedback not only inflates confidence, thereby inducing overconfidence, but also harms the confidence-accuracy correlation. When eyewitnesses are given confirming feedback following their identification decisions, the confidence of inaccurate eyewitnesses is inflated more than is the confidence of accurate eyewitnesses, and the net result is a reduction in the confidence-accuracy correlation (Bradfield, Wells, & Olson, 2002). Hence, although the confidence of an eyewitness can have utility if it is assessed independently of external influences (e.g., comments from the detective, learning about what other eyewitnesses have said), the legal system rarely assesses confidence in this way.

IMPACT ON POLICIES AND PRACTICES

What impact has research on the confidence-accuracy problem had on the legal system? Until relatively recently, the impact has been almost nil. However, when DNA exoneration cases began unfolding in the mid-1990s, U.S. Attorney General Janet Reno initiated a study of the causes of these miscarriages of justice. More than three fourths of these convictions of innocent persons involved mistaken eyewitness identifications, and, in every case, the mistaken eyewitnesses were extremely confident and, therefore, persuasive at trial (Wells et al., 1998). A Department of Justice panel used the psychological literature to issue the first set of national guidelines on collecting eyewitness identification evidence (Technical Working Group for Eyewitness Evidence, 1999). One of the major recommendations was that the confidence of the eyewitness be assessed at the time of the identification, before there is any chance for it to be influenced by external factors.

The state of New Jersey has gone even further in adopting the recommendations of eyewitness researchers. Based on findings from the psychological literature, guidelines from the attorney general of New Jersey now call for double-blind testing with lineups. Double-blind lineup testing means that the person who administers the lineup does not know which person in the lineup is the suspect and which ones are merely fillers. Under the New Jersey procedures, the confidence expressed by the eyewitness will be based primarily on the eyewitness's memory, not on the expectations of or feedback from the lineup administrator.

There is growing evidence that the legal system is now beginning to read and use the psychological literature on eyewitnesses to formulate policies and procedures. The 2002 report of Illinois Governor George Ryan's Commission on Capital Punishment is the latest example of this new reliance on the psychological literature. The commission specifically cited the literature on the problem with confidence inflation and recommended double-blind testing and explicit recording of confidence statements at the time of the identification to prevent or detect confidence inflation (Illinois Commission on Capital Punishment, 2002).

NEW DIRECTIONS

Although the psychological literature on eyewitness identification has done much to clarify the confidence-accuracy issue and specify some conditions under which confidence might be predictive of accuracy, research has started to turn to other indicators that might prove even more predictive of accuracy. One of the most promising examples is the relation between the amount of time an eyewitness takes to make an identification and the accuracy of the identification. Eyewitnesses who make their identification decision quickly (in 10 s or less) are considerably more likely to be accurate than are eyewitnesses who take longer (e.g., Dunning & Perretta, in press). Confidence is a self-report that is subject to distortion (e.g., from postidentification feedback), whereas decision time is a behavior that can be directly observed. Hence, decision time might prove more reliable than confidence as an indicator of eyewitness accuracy. Yet another new direction in eyewitness identification research concerns cases in which there are multiple eyewitnesses. Recent analyses show that the behaviors of eyewitnesses who do not identify the suspect from a lineup can be used to assess the likely accuracy of the eyewitnesses who do identify the suspect from a lineup (Wells & Olson, in press). The future of eyewitness identification research is a bright one, and the legal system now seems to be paying attention.

Recommended Reading

Cutler, B.L., & Penrod, S.D. (1995). *Mistaken identification: The eyewitness, psychology, and the law.* New York: Cambridge University Press.

Scheck, B., Neufeld, P., & Dwyer, J. (2000). (See References)

Wells, G.L., Malpass, R.S., Lindsay, R.C.L., Fisher, R.P., Turtle, J.W., & Fulero, S. (2000). From the lab to the police station: A successful application of eyewitness research. *American Psychologist*, 55, 581–598.

Note

1. Address correspondence to Gary L. Wells, Psychology Department, Iowa State University, Ames, IA 50011; e-mail: glwells@iastate.edu.

References

Bradfield, S.L., Wells, G.L., & Olson, E.A. (2002). The damaging effect of confirming feedback on the relation between eyewitness certainty and identification accuracy. *Journal of Applied Psychology*, 87, 112–120.

Brewer, N., Keast, A., & Rishworth, A. (2002). Improving the confidence-accuracy relation in eyewitness identification: Evidence from correlation and calibration. *Journal of Experimental Psychology: Applied, 8*, 44–56.
Department of Health and Human Services, National Center for Health Statistics. (n.d.). *National Health and Nutrition Examination Survey.* Retrieved May 22, 2002, from http://www.cdc.gov/nchs/about/major/nhanes/datatblelink.htm#unpubtab
Dunning, D., & Perretta, S. (in press). Automaticity and eyewitness accuracy: A 10 - to - 12 second rule for distinguishing accurate from inaccurate positive identifications. *Journal of Applied Psychology.*
Illinois Commission on Capital Punishment. (2002, April). *Report of the Governor's Commission on Capital Punishment.* Retrieved April 23, 2002, from http://www.idoc.state.il.us/ccp/ccp/reports/commission_report/
Juslin, P., Olson, N., & Winman, A. (1996). Calibration and diagnosticity of confidence in eyewitness identification: Comments on what can and cannot be inferred from a low confidence-accuracy correlation. *Journal of Experimental Psychology: Learning, Memory, and Cognition, 5*, 1304–1316.
Read, J.D., Vokey, J.R., & Hammersley, R. (1990). Changing photos of faces: Effects of exposure duration and photo similarity on recognition and the accuracy-confidence relationship. *Journal of Experimental Psychology: Learning, Memory, and Cognition, 16*, 870–882.
Scheck, B., Neufeld, P., & Dwyer, J. (2000). *Actual innocence.* New York: Random House.
Shaw, J.S., III. (1996). Increases in eyewitness confidence resulting from postevent questioning. *Journal of Experimental Psychology: Applied, 2*, 126–146.
Shaw, J.S., III, & McClure, K.A. (1996). Repeated postevent questioning can lead to elevated levels of eyewitness confidence. *Law and Human Behavior, 20*, 629–654.
Sporer, S., Penrod, S., Read, D., & Cutler, B.L. (1995). Choosing, confidence, and accuracy: A meta-analysis of the confidence-accuracy relation in eyewitness identification studies. *Psychological Bulletin, 118*, 315–327.
Technical Working Group for Eyewitness Evidence. (1999). *Eyewitness evidence: A guide for law enforcement.* Washington, DC: U.S. Department of Justice, Office of Justice Programs.
Wells, G.L., & Bradfield, A.L. (1999). Distortions in eyewitnesses' recollections: Can the postidentification-feedback effect be moderated? *Psychological Science, 10*, 138–144.
Wells, G.L., & Olson, E.A. (in press). Eyewitness identification: Information gain from incriminating and exonerating behaviors. *Journal of Experimental Psychology: Applied.*
Wells, G.L., Small, M., Penrod, S., Malpass, R.S., Fulero, S.M., & Brimacombe, C.A.E. (1998). Eyewitness identification procedures: Recommendations for lineups and photospreads. *Law and Human Behavior, 22*, 603–647.

Critical Thinking Questions

1. What factors drive confidence and accuracy in the same direction? What factors might drive one but not the other? Which of these occur at the crime scene (and are not under anyone's direct control) and which are the result of post-crime procedures (and are controllable)?

2. The article mentions that repeated questioning of a witness can lead to increases in confidence (but not accuracy). How does that finding remind you of the issues in the Garry and Polaschek article?

3. Two types of errors can be made by a witness when viewing a line-up: the witness might fail to identify the actual criminal or the witness might improperly identify an innocent suspect. Are the consequences of those two types of error equal? With which type of error is this article more concerned? Why?

4. Current law tells jurors that they may consider witness confidence as relevant to evaluating the witness's accuracy and credibility. Do you believe that jurors should be given instructions regarding the *value* of a witness's confidence on the witness stand? What do you think those instructions should be?

5. In the last sentence, the authors state that "the legal system now seems to be paying attention" to eyewitness identification research. Good research in this area has been around for many years. Why do you think they are paying attention to this research "now"? Is the legal system all of a sudden open to using all social science research or is this a fluke (or maybe a foot in the door)?

6. Suppose you were a witness to a crime. What might you want to do (or not do) to preserve the accuracy of your knowledge/memory? What might you want to do (or not do) to make sure your confidence level reflects your accuracy?

Making Group Brainstorming More Effective: Recommendations From an Associative Memory Perspective

Vincent R. Brown and Paul B. Paulus[1]

Department of Psychology, Hofstra University, Hempstead, New York (V.R.B.), and Department of Psychology, University of Texas at Arlington, Arlington, Texas (P.B.P.)

Abstract

Much literature on group brainstorming has found it to be less effective than individual brainstorming. However, a cognitive perspective suggests that group brainstorming could be an effective technique for generating creative ideas. Computer simulations of an associative memory model of idea generation in groups suggest that groups have the potential to generate ideas that individuals brainstorming alone are less likely to generate. Exchanging ideas by means of writing or computers, alternating solitary and group brainstorming, and using heterogeneous groups appear to be useful approaches for enhancing group brainstorming.

Keywords

brainstorming; cognitive stimulation; groups; group creativity

There is a general belief in the efficacy of collaboration for projects involving innovation or problem solving (Bennis & Biederman, 1997; Sutton & Hargadon, 1996). Although there is some evidence for the effectiveness of collaborative science and teamwork (Paulus, 2000), the enthusiasm for collective work may not always be justified. Controlled studies of idea sharing in groups have shown that groups often overestimate their effectiveness (Paulus, Larey, & Ortega, 1995). Experiments comparing interactive brainstorming groups with sets of individuals who do not interact in performing the same task have found that groups generate fewer ideas and that group members exhibit reduced motivation and do not fully share unique information (e.g., Mullen, Johnson, & Salas, 1991). The strongest inhibitory effect of groups may be production blocking, which is a reduction in productivity due to the fact that group members must take turns in describing their ideas (Diehl & Stroebe, 1991).

One area in which these problems are most evident is the study of group creativity. Most research on creativity has examined individual creativity because it is typically seen as a personal trait or skill. However, today much creative work requires collaboration of people with diverse sets of knowledge and skills. How can such groups overcome the inevitable liabilities of group interaction to reach their creative potential? Is it possible to demonstrate that group interaction can lead to enhanced creativity? Examining these questions has been the aim of our program of research on the cognitive potential of brainstorming groups (Brown, Tumeo, Larey, & Paulus, 1998; Paulus & Brown, in press; Paulus, Dugosh, Dzindolet, Coskun, & Putman, 2002).

COGNITIVE BASES FOR IDEATIONAL CREATIVITY: A MODEL AND SUPPORTING EVIDENCE

Intuitively, the cognitive benefits of brainstorming in a group seem clear: People believe that they come up with ideas in a group that they would not have thought of on their own. The potential for mutual stimulation of ideas is one of the reasons for the popularity of group brainstorming.

Semantic Networks and an Associative Memory Model of Group Brainstorming

Clearly, the retrieval of relevant information from one's long-term conceptual memory is an important part of the brainstorming process because one cannot effectively brainstorm on a topic one knows nothing about. The concepts stored in long-term memory can be thought of as being associated in a semantic network in such a way that related concepts are more strongly connected than unrelated concepts and thus more likely to activate each other (e.g., Collins & Loftus, 1975). Thus, concepts that are more closely connected to those that are currently active should be more accessible than concepts that are less strongly connected to currently active ideas.

To use the semantic network representation as a basis for exploring group brainstorming, many details need to be specified. Rather than explicitly representing four, six, eight, or more semantic networks and the interactions among them—which would be cumbersome—our approach is to represent a brainstormer's knowledge of a given problem as a matrix of category transition probabilities: Each entry in this matrix represents the probability of generating one's next idea from the same category as the previous idea or from a different category (Brown et al., 1998).

A number of individual differences affecting brainstorming performance are captured by this framework. Fluency, or the amount of knowledge one has about the brainstorming problem, is represented by the probabilities in the main body of the matrix relative to the null category, which represents the likelihood of coming up with no idea in a given time interval. Convergent and divergent thinking styles also fit nicely into the framework. On the one hand, a convergent thinker is likely to stick with a category and explore it deeply before moving on to generate ideas from other categories. Thus, a convergent thinker is represented by a matrix with relatively high within-category transition probabilities. On the other hand, a divergent thinker is more likely to jump around between categories, and so is represented by a matrix with lower within-category transition probabilities (and correspondingly higher between-category transition probabilities). Because individual differences can be represented in this framework, the effects of different group compositions can be examined.

Accessibility

The property of the individual's semantic network that is crucial to determining the effectiveness of group brainstorming is category accessibility. People are generally unlikely to explore on their own relevant categories of ideas that have relatively weak connections with other categories in their personal semantic networks. Generation of ideas from these categories requires the spark of input from other brainstormers. For example, a student who lives on campus in a dormitory may be unlikely to generate ideas about parking when brainstorming on ways to improve the university. But if a student who commutes from off campus mentions parking, the dorm dweller may be able to come up with a few thoughts on the matter, perhaps recalling the parking difficulties his or her parents had when they visited. Simulations of the associative memory matrix model show that presenting a brainstormer with ideas from low-accessible categories not only increases the number of ideas generated from those categories, but also increases the total number of ideas generated overall, thus making the individual a more productive brainstormer. This prediction is supported by Leggett (1997), who studied individual brainstormers to evaluate how input of ideas in the absence of a group context (i.e., without inhibitory social influences) might provide cognitive stimulation. Participants listened to audiotapes containing ideas generated by participants who worked on the Thumbs Problem ("What would be the advantages and disadvantages of having an extra thumb on each hand?") in previous studies. Although all brainstormers benefited from being primed with relevant ideas, the amount of benefit depended on whether the ideas came from a category that was frequently or infrequently represented in the responses in the previous studies. Individuals who were primed with ideas from common categories obtained less benefit than those who were primed with ideas from unique categories. In other words, priming categories that were already likely to be utilized did not enhance performance as much as priming categories unlikely to be utilized by someone brainstorming on his or her own.

Attention

Individuals will be influenced by other group members to the extent that they pay attention to those other members' ideas. In the framework of the model, attention is represented as the probability that an individual group member uses the current speaker's idea as the basis for generating his or her next idea (as opposed to simply continuing his or her own internal train of thought).

Simulations predict that, in general, the more attention each individual pays to fellow group members, the better the performance of the group. Conversely, the more each individual's attention is distracted from the ideas of others, perhaps by concern for the social aspects of group brainstorming, the more the performance of the group declines. In particular, the more one attends to fellow brainstormers, the more one is likely to be primed to consider ideas from one's own low-accessible categories. In fact, the model predicts that, in general, if it were not for production blocking, the number of ideas generated by each group member would increase (at least up to a point) as group size increases (Brown et al., 1998).

One way to enhance the effects of attention on brainstorming performance is to instruct brainstormers that at the end of the brainstorming session they will be asked to recall the ideas that were presented during the session. Without memorization instructions, participants may be more likely to focus on the generation of their own ideas and to some extent ignore the ideas being presented by others. Interestingly, the effectiveness of memory instructions appears to be mixed. When participants are listening to audiotapes or exchanging ideas by computer, instructions to memorize facilitate idea generation (Dugosh, Paulus, Roland, & Yang, 2000). However, when participants are asked to exchange written ideas in a round-robin format (Paulus & Yang, 2000), memory instructions inhibit performance. Because the instructions to read the ideas as they are passed from person to person may already ensure that participants attend to the ideas, instructions to memorize may simply add an unnecessary additional processing demand and impede the brainstorming effort.

ENHANCING GROUP BRAINSTORMING

The goal of creating circumstances that optimize group performance requires maximizing the benefits of cognitive facilitation while at the same time minimizing the inhibitory processes that reduce group productivity. We have studied three brainstorming procedures that appear promising for theoretical reasons, and that have garnered some empirical support. These are combining group and solitary brainstorming, having group brainstormers interact by writing instead of speaking ("brainwriting"), and using networked computers on which individuals type their ideas and read the ideas of others (electronic brainstorming).

Individual and Group Brainstorming

At face value, the goal of maximizing the benefits of group exchange while minimizing inhibitory group processes suggests literally combining group and individual brainstorming. Of course, a person cannot brainstorm alone while at the same time brainstorming in a group. But one can alternate group and solitary idea-generation sessions. Preliminary data from our laboratory show that brainstorming in a group before brainstorming alone on the same topic produces more ideas over the course of the two sessions than does brainstorming alone in the first session and then brainstorming in a group the following session (Leggett, Putman, Roland, & Paulus, 1996).

Model simulations make clear the mechanisms that produce this advantage for the group-solitary sequence. The cognitive facilitation that occurs in the group session carries over into the solitary session, during which the brainstormer continues generating ideas without being hindered by production blocking. This effect shows up as a large "productivity spike" for solitary brainstormers in the second session in both model simulations and empirical studies. This order effect should be particularly strong when the initial group consists of heterogeneous members whose knowledge of the task differs. Simulations also indicate that a solitary brainstormer whose idea generation takes place following a group session is likely to sample more categories from the brainstorming topic than a similar brainstormer working in two solitary sessions. This suggests that the group-solitary sequence has an advantage over and above possible increases in overall productivity.

Brainwriting

Another way to take advantage of group priming effects while reducing production blocking would be to have group members interact by writing and reading rather than speaking and listening. This does not seem to be a technique that is often attempted. Perhaps people are so used to communicating orally when they are face to face that researchers do not consider the alternatives. In a study of group brainwriting (Paulus & Yang, 2000), group members wrote their ideas on a piece of paper and passed them on to the next group member, who read the ideas, added his or her own ideas, and passed the paper on. These interactive groups of brainwriters outproduced sets of equal numbers of writers who did not interact in performing the task. This result may be the first laboratory example of face-to-face interactive groups outperforming an equal number of solitary brainstormers.

Although model simulations support the observation that interactive brainwriters can outperform an equal number of solitary brainwriters (Paulus & Brown, in press), the simulation results are complex in some interesting ways. First, simulations predict that interactive brainwriting is not universally superior to individual brainwriting, but is most effective for heterogeneous groups whose members have differing knowledge of the brainstorming problem. Second, up to a point, performance of simulated brainwriting groups improves as the group members pay increasing attention to the written ideas, but performance decreases when attention to the written ideas becomes excessive. Obviously, brainwriters who do not read any of the ideas that are passed along to them will not benefit from the thoughts of their fellow brainwriters. Brainwriters who attend predominantly to the ideas of others will benefit from them to some extent, but not as greatly as those who optimally balance the two goals of attending to the ideas of others and following their own internal train of thought.

Electronic Brainstorming

The effectiveness of brainwriting suggests that using computers for exchanging ideas might be another useful way of tapping the creative potential of groups. Using computers, individuals can be exposed to a broad range of ideas without the verbal "traffic jams" that are problematic for face-to-face groups. In fact, because production blocking is greatly reduced, the larger the group the better—larger groups will increase one's exposure to a broad range of ideas. One of the most consistent findings in the electronic-brainstorming literature is that the benefits are most evident for groups of eight or more (Dennis & Williams, in press).

Unfortunately, it is also easier to ignore the inputs of others in the electronic format than in face-to-face situations. Instructing individuals to attend carefully to ideas shared electronically (e.g., because of an impending memory test) increases the impact of the shared information (Dugosh et al., 2000). Although electronic brainstormers do not have to apply extensive intellectual resources to try to remember the shared ideas because they are available on the computer, for large groups the number of available ideas can become rather overwhelming. It may be important to provide an opportunity for individuals to continue processing the ideas after the interactive session in order to gain full associative advantage of the shared information. The benefits of idea sharing in electronic groups are in fact most evident if individuals are provided such a solitary session after group interaction.

CONCLUSIONS

It is clear that unstructured groups left to their own devices will not be very effective in developing creative ideas. However, a cognitive perspective points to methods that can be used so that group exchange of ideas enhances idea generation. Groups of individuals with diverse sets of knowledge are most likely to benefit from the social exchange of ideas. Although face-to-face interaction is seen as a natural modality for group interaction, using writing or computers can enhance the exchange of ideas. The interaction should be structured to ensure careful attention to the shared ideas. Alternating between individual and group ideation is helpful because it allows for careful reflection on and processing of shared ideas.

There are still a number of significant empirical gaps that need to be addressed. Given that much group exchange consists of verbal interaction in face-to-face groups, studies need to determine the specific extent to which the performance of these groups can be enhanced by using insights from the associative memory perspective. In particular, it will be important to demonstrate that groups that contain members with diverse knowledge bases can effectively use this knowledge interactively for creative purposes. There are also no controlled studies of creativity in groups or teams in organizations outside the laboratory, so that it is not possible to draw definitive conclusions about the effectiveness of groups in the real world. Because group interaction can be a source of social and cognitive interference as well as social and cognitive stimulation, one of the main theoretical challenges will be to integrate the cognitive and social perspectives of group brainstorming. A careful delineation of how these processes interact will be of great benefit to practitioners.

Recommended Reading

Brown, V., Tumeo, M., Larey, T.S., & Paulus, P.B. (1998). (See References)
Osborn, A.F. (1957). *Applied imagination* (1st ed.). New York: Scribner.
Paulus, P.B., & Nijstad, B.A. (Eds.). (in press). *Group creativity*. New York: Oxford University Press.

Note

1. Address correspondence to Paul B. Paulus, Department of Psychology, University of Texas at Arlington, Arlington, TX 76019; e-mail: paulus@uta. edu.

References

Bennis, W., & Biederman, P.W. (1997). *Organizing genius: The secrets of creative collaboration*. Reading, MA: Addison Wesley.

Brown, V., Tumeo, M., Larey, T.S., & Paulus, P.B. (1998). Modeling cognitive interactions during group brainstorming. *Small Group Research, 29*, 495–526.

Collins, A.M., & Loftus, E.F. (1975). A spreading-activation theory of semantic processing. *Psychological Review, 82*, 407–428.

Dennis, A.R., & Williams, M.L. (in press). Electronic brainstorming: Theory, research, and future directions. In P.B. Paulus & B.A. Nijstad (Eds.), *Group creativity.* New York: Oxford University Press.

Diehl, M., & Stroebe, W. (1991). Productivity loss in idea-generating groups: Tracking down the blocking effect. *Journal of Personality and Social Psychology, 61*, 392–403.

Dugosh, K.L., Paulus, P.B., Roland, E.J., & Yang, H.-C. (2000). Cognitive stimulation in brainstorming. *Journal of Personality and Social Psychology, 79*, 722–735.

Leggett, K.L. (1997). *The effectiveness of categorical priming in brainstorming.* Unpublished master's thesis, University of Texas at Arlington.

Leggett, K.L., Putman, V.L., Roland, E.J., & Paulus, P.B. (1996, April). *The effects of training on performance in group brainstorming.* Paper presented at the annual meeting of the Southwestern Psychological Association, Houston, TX.

Mullen, B., Johnson, C., & Salas, E. (1991). Productivity loss in brainstorming groups: A meta-analytic integration. *Basic and Applied Social Psychology, 12*, 3–23.

Paulus, P.B. (2000). Groups, teams and creativity: The creative potential of idea generating groups. *Applied Psychology: An International Review, 49*, 237–262.

Paulus, P.B., & Brown, V. (in press). Ideational creativity in groups: Lessons from research on brainstorming. In P.B. Paulus & B.A. Nijstad (Eds.), *Group creativity.* New York: Oxford University Press.

Paulus, P.B., Dugosh, K.L., Dzindolet, M.T., Coskun, H., & Putman, V.L. (2002). Social and cognitive influences in group brainstorming: Predicting production gains and losses. *European Review of Social Psychology, 12*, 299–325.

Paulus, P.B., Larey, T.S., & Ortega, A.H. (1995). Performance and perceptions of brainstormers in an organizational setting. *Basic and Applied Social Psychology, 18*, 3–14.

Paulus, P.B., & Yang, H.-C. (2000). Idea generation in groups: A basis for creativity in organizations. *Organizational Behavior and Human Decision Processes, 82*, 76–87.

Sutton, R.I., & Hargadon, A. (1996). Brainstorming groups in context. *Administrative Science Quarterly, 41*, 685–718.

Critical Thinking Questions

1. This article relies on the notion of memory as an associative network. What other memory effects—within individual, not group, memory—have you studied that are easily explainable in terms of semantic or associative networks?

2. This article is reminiscent of work on part-set cuing. In part-set cuing studies, subjects may be asked to write down all of the names of U. S. states that they can think of. The "cued" subjects get a list of several states in front of them to "help them start"; the "non-cued" subjects do not. The interesting finding is that non-cued subjects remember more of the unlisted states than the cued subjects. One explanation is that the list keeps bringing the cued subjects' thoughts back to the same items. Those subjects then repeatedly recollect the same states, and the states they have mentally stored as associated to those states, but they can't break out of those mental loops. How can the techniques of group brainstorming solve that problem? Why might it depend on the method of information-sharing used by the group?

3. Why might people working in groups overestimate their own effectiveness in brainstorming more than individuals working alone?

4. Suppose you were in charge of designing an advertising campaign for a new type of product. You want to select several people from your company to get together for a brainstorming session. Who do you pick? A bunch of people who have worked together many times before and who get along well? People from different previous teams who have not worked together before and don't personally like each other? What are the cognitive and social trade-offs to each of these selections?

   (How and why would your personnel selection for this task differ from your personnel selection if you were seeking to evaluate a product or a marketing strategy? See Yaniv in Part 4.)

Associative Learning and Causal Reasoning

An essential skill for survival is being able to learn about contingencies in the world. Humans have learned which berries are edible, which metals make good weapons, and which medicines cure diseases. How do we pick up such knowledge? Do we all always use it accurately—or similarly? The four articles in this section look at different factors that may affect such learning and reasoning including: evolutionary pre-dispositions, personality traits, imagination, and culture.

The first two articles consider what might be called "associative learning". They describe studies in which subjects (sometimes human, sometimes other species) are exposed over and over to some stimulus; later a response is measured. In "The Malicious Serpent: Snakes as a Prototypical Stimulus for an Evolved Module of Fear," Öhman and Mineka describe both classical conditioning and illusory correlation paradigms for examining fear reactions to snakes. In each paradigm, subjects are shown pictures of various items (e.g., snakes, flowers) that are sometimes paired with an unpleasant shock. In the conditioning paradigm, subjects exhibit stronger fear conditioning to snakes than to flowers. In the illusory correlation paradigm, subjects believe that the snakes were paired more often with the shocks than were the flowers (even when the proportions of such pairings were equal). The authors provide an evolutionary account for why such differential learning might take place.

In "Illusions of Control: How We Overestimate Our Personal Influence," Thompson describes several operant conditioning experiments in which subjects are led to believe that they have control over some outcome (e.g., if they push a button a light will turn on). The outcome may occur totally by chance or it may be a function of both the subjects' actions and chance. When asked to judge how much control they have over the outcome, subjects often overestimate their own causal power. Thompson suggests which kinds of people are most likely to overestimate their causal influence on events and when they are most likely to do so. She ends by speculating on the adaptiveness of the illusory control bias in judgment.

The next two articles reflect on how people determine causality for one-time events—like a car accident or a murder. In "When Possibility Informs Reality: Counterfactual Thinking as a Cue to Causality," Spellman and Mandel consider how counterfactual reasoning might affect a person's causal reasoning about such events. In counterfactual reasoning, we imagine the world different from how it actually is (i.e., counter-to-fact) and play out the consequences. For example, you might imagine that if Timothy McVeigh (the Oklahoma City bomber) had had a different upbringing, he would not have blown up the building. The article describes the controversy over whether such a counterfactual relationship means it is correct to say that the bad upbringing *caused* him to blow up the building.

Whether people will assign causality for a person's behavior to a disposition (i.e., a personal trait) or a situation (i.e., the circumstances) is the topic addressed by Norenzayan and Nisbett in "Culture and Causal Cognition". Whereas Westerners tend to assign causality to an individual's traits (he was always a destructive child), East Asians tend to assign causality to a situation (he was always pressured by others to do so). Thus, different people may make different causal attributions for the same set of events.

All four articles demonstrate, therefore, that our judgments about the relationship between events in the world are not solely based on the contingencies in the world. Rather, our causal judgments are influenced by pre-existing knowledge and beliefs—both conscious and unconscious, both innate and learned.

The Malicious Serpent: Snakes as a Prototypical Stimulus for an Evolved Module of Fear

Arne Öhman[1] and Susan Mineka
Department of Clinical Neuroscience, Karolinska Institute, Stockholm, Sweden (A.Ö.), and Department of Psychology, Northwestern University, Evanston, Illinois (S.M.)

Abstract

As reptiles, snakes may have signified deadly threats in the environment of early mammals. We review findings suggesting that snakes remain special stimuli for humans. Intense snake fear is prevalent in both humans and other primates. Humans and monkeys learn snake fear more easily than fear of most other stimuli through direct or vicarious conditioning. Neither the elicitation nor the conditioning of snake fear in humans requires that snakes be consciously perceived; rather, both processes can occur with masked stimuli. Humans tend to perceive illusory correlations between snakes and aversive stimuli, and their attention is automatically captured by snakes in complex visual displays. Together, these and other findings delineate an evolved fear module in the brain. This module is selectively and automatically activated by once-threatening stimuli, is relatively encapsulated from cognition, and derives from specialized neural circuitry.

Keywords

evolution; snake fear; fear module

Snakes are commonly regarded as shiny, slithering creatures worthy of fear and disgust. If one were to believe the Book of Genesis, humans' dislike for snakes resulted from a divine intervention: To avenge the snake's luring of Eve to taste the fruit of knowledge, God instituted eternal enmity between their descendants. Alternatively, the human dislike of snakes and the common appearances of reptiles as the embodiment of evil in myths and art might reflect an evolutionary heritage. Indeed, Sagan (1977) speculated that human fear of snakes and other reptiles may be a distant effect of the conditions under which early mammals evolved. In the world they inhabited, the animal kingdom was dominated by awesome reptiles, the dinosaurs, and so a prerequisite for early mammals to deliver genes to future generations was to avoid getting caught in the fangs of Tyrannosaurus rex and its relatives. Thus, fear and respect for reptiles is a likely core mammalian heritage. From this perspective, snakes and other reptiles may continue to have a special psychological significance even for humans, and considerable evidence suggests this is indeed true. Furthermore, the pattern of findings appears consistent with the evolutionary premise.

THE PREVALENCE OF SNAKE FEARS IN PRIMATES

Snakes are obviously fearsome creatures to many humans. Agras, Sylvester, and Oliveau (1969) interviewed a sample of New Englanders about fears, and found snakes to be clearly the most prevalent object of intense fear, reported by 38% of females and 12% of males.

Fear of snakes is also common among other primates. According to an exhaustive review of field data (King, 1997), 11 genera of primates showed fear-related responses (alarm calls, avoidance, mobbing) in virtually all instances in which they were observed confronting large snakes. For studies of captive primates, King did not find consistent evidence of snake fear. However, in direct comparisons, rhesus (and squirrel) monkeys reared in the wild were far more likely than lab-reared monkeys to show strong phobiclike fear responses to snakes (e.g., Mineka, Keir, & Price, 1980). That this fear is adaptive in the wild is further supported by independent field reports of large snakes attacking primates (M. Cook & Mineka, 1991).

This high prevalence of snake fear in humans as well as in our primate relatives suggests that it is a result of an ancient evolutionary history. Genetic variability might explain why not all individuals show fear of snakes. Alternatively, the variability could stem from differences in how easily individuals learn to fear reptilian stimuli when they are encountered in aversive contexts. This latter possibility would be consistent with the differences in snake fear between wild- and lab-reared monkeys.

LEARNING TO FEAR SNAKES

Experiments with lab-reared monkeys have shown that they can acquire a fear of snakes vicariously, that is, by observing other monkeys expressing fear of snakes. When nonfearful lab-reared monkeys were given the opportunity to observe a wild-reared "model" monkey displaying fear of live and toy snakes, they were rapidly conditioned to fear snakes, and this conditioning was strong and persistent. The fear response was learned even when the fearful model monkey was shown on videotape (M. Cook & Mineka, 1990).

When videos were spliced so that identical displays of fear were modeled in response to toy snakes and flowers, or to toy crocodiles and rabbits (M. Cook & Mineka, 1991), the lab-reared monkeys showed substantial conditioning to toy snakes and crocodiles, but not to flowers and toy rabbits. Toy snakes and flowers served equally well as signals for food rewards (M. Cook & Mineka, 1990), so the selective effect of snakes appears to be restricted to aversive contexts. Because these monkeys had never seen any of the stimuli used prior to these experiments, the results provide strong support for an evolutionary basis to the selective learning.

A series of studies published in the 1970s (see Öhman & Mineka, 2001) tested the hypothesis that humans are predisposed to easily learn to fear snakes. These studies used a discriminative Pavlovian conditioning procedure in which various pictures served as conditioned stimuli (CSs) that predicted the presence and absence of mildly aversive shock, the unconditioned stimulus (US). Participants for whom snakes (or spiders) consistently signaled shocks showed stronger and more lasting conditioned skin conductance responses (SCRs; palmar sweat responses that index emotional activation) than control participants for whom flowers or mushrooms signaled shocks. When a nonaversive US was used, however, this difference disappeared. E.W. Cook, Hodes, and Lang (1986) demonstrated that qualitatively different responses were conditioned to snakes (heart rate acceleration, indexing fear) than to flowers and mushrooms (heart rate deceleration, indexing attention to the eliciting stimulus). They also reported superior conditioning to snakes than to gun stimuli paired with loud noises. Such results suggest that the selective association between snakes and aversive USs reflects evolutionary history rather than cultural conditioning.

NONCONSCIOUS CONTROL OF RESPONSES TO SNAKES

If the prevalence and ease of learning snake fear represents a core mammalian heritage, its neural machinery must be found in brain structures that evolved in early mammals. Accordingly, the fear circuit of the mammalian brain relies heavily on limbic structures such as the amygdala, a collection of neural nuclei in the anterior temporal lobe. Limbic structures emerged in the evolutionary transition from reptiles to mammals and use preexisting structures in the "reptilian brain" to control emotional output such as flight/fight behavior and cardiovascular changes (see Öhman & Mineka, 2001).

From this neuroevolutionary perspective, one would expect the limbically controlled fear of snakes to be relatively independent of the most recently evolved control level in the brain, the neocortex, which is the site of advanced cognition. This hypothesis is consistent with the often strikingly irrational quality of snake phobia. For example, phobias may be activated by seeing mere pictures of snakes. Backward masking is a promising methodology for examining whether phobic responses can be activated without involvement of the cortex. In this method, a brief visual stimulus is blanked from conscious perception by an immediately following masking stimulus. Because backward masking disrupts visual processing in the primary visual cortex, responses to backward-masked stimuli reflect activation of pathways in the brain that may access the fear circuit without involving cortical areas mediating visual awareness of the stimulus.

In one study (Öhman & Soares, 1994), pictures of snakes, spiders, flowers, and mushrooms were presented very briefly (30 ms), each time immediately followed by a masking stimulus (a randomly cut and reassembled picture). Although the participants could not recognize the intact pictures, participants who were afraid of snakes showed enhanced SCRs only to masked snakes, whereas participants who were afraid of spiders responded only to spiders. Similar results were obtained (Öhman & Soares, 1993) when nonfearful participants, who had been conditioned to unmasked snake pictures by shock USs, were exposed to masked pictures without the US. Thus, responses to conditioned snake pictures survived backward masking; in contrast, masking eliminated conditioning effects in another group of participants conditioned to neutral stimuli such as flowers or mushrooms.

Furthermore, subsequent experiments (Öhman & Soares, 1998) also demonstrated conditioning to masked stimuli when masked snakes or spiders (but not masked flowers or mushrooms) were used as CSs followed by shock USs. Thus, these masking studies show that fear responses (as indexed by SCRs) can be learned and elicited when backward masking prevents visually presented snake stimuli from accessing cortical processing. This is consistent with the notion that responses to snakes are organized by a specifically evolved primitive neural circuit that emerged with the first mammals long before the evolution of neocortex.

ILLUSORY CORRELATIONS BETWEEN SNAKES AND AVERSIVE STIMULI

If expression and learning of snake fear do not require cortical processing, are people's cognitions about snakes and their relationships to other events biased and irrational? One example of such biased processing occurred in experiments on illusory correlations: Participants (especially those who were afraid of snakes) were more likely to perceive that slides of fear-relevant stimuli (such as snakes) were paired with shock than to perceive that slides of control stimuli (flowers and mushrooms) were paired with shock. This occurred even though there were no such relationships in the extensive random sequence of slide stimuli and aversive and nonaversive outcomes (tones or nothing) participants had experienced (Tomarken, Sutton, & Mineka, 1995).

Similar illusory correlations were not observed for pictures of damaged electrical equipment and shock even though they were rated as belonging together better than snakes and shock (Tomarken et al., 1995). In another experiment, participants showed exaggerated expectancies for shock to follow both snakes and damaged electrical equipment before the experiment began (Kennedy, Rapee, & Mazurski, 1997), but reported only the illusory correlation between snakes and shock after experiencing the random stimulus series. Thus, it appears that snakes have a cognitive affinity with aversiveness and danger that is resistant to modification by experience.

AUTOMATIC CAPTURE OF ATTENTION BY SNAKE STIMULI

People who encounter snakes in the wild may report that they first froze in fear, only a split second later realizing that they were about to step on a snake. Thus, snakes may automatically capture attention. A study supporting this hypothesis (Öhman, Flykt, & Esteves, 2001) demonstrated shorter detection latencies for a discrepant snake picture among an array of many neutral distractor stimuli (e.g., flower pictures) than vice versa. Furthermore, "finding the snake in the grass" was not affected by the number of distractor stimuli, whereas it took longer to detect discrepant flowers and mushrooms among many than among few snakes when the latter served as distractor stimuli. This suggests that snakes, but not flowers and mushrooms, were located by an automatic perceptual routine that effortlessly found target stimuli that appeared to "pop out" from the matrix independently of the number of distractor stimuli. Participants who were highly fearful of snakes showed even superior performance in detecting snakes. Thus, when snakes elicited fear in participants, this fear state sensitized the perceptual apparatus to detect snakes even more efficiently.

THE CONCEPT OF A FEAR MODULE

The evidence we have reviewed shows that snake stimuli are strongly and widely associated with fear in humans and other primates and that fear of snakes is relatively independent of conscious cognition. We have proposed the concept of an evolved fear module to explain these and many related findings (Öhman & Mineka, 2001). The fear module is a relatively independent behavioral, mental, and neural system that has evolved to assist mammals in defending against threats such as snakes. The module is selectively sensitive to, and automatically activated by, stimuli related to recurrent survival threats, it is relatively encapsulated from more advanced human cognition, and it relies on specialized neural circuitry.

This specialized behavioral module did not evolve primarily from survival threats provided by snakes during human evolution, but rather from the threat that reptiles have provided through mammalian evolution. Because reptiles have been associated with danger throughout evolution, it is likely that snakes represent a prototypical stimulus for activating the fear module. However, we are not arguing that the human brain has a specialized module for automatically generating fear of snakes. Rather, we propose that the blueprint for the fear module was built around the deadly threat that ancestors of snakes provided to our distant ancestors, the early mammals. During further mammalian evolution, this blueprint was modified, elaborated, and specialized for the ecological niches occupied by different species. Some mammals may even prey on snakes, and new stimuli and stimulus features have been added to reptiles as preferential activators of the module. For example, facial threat is similar to snakes when it comes to activating the fear module in social primates (Öhman & Mineka, 2001). Through Pavlovian conditioning, the fear module may come under the control of a very wide range of stimuli signaling pain and danger. Nevertheless, evolutionarily derived constraints have afforded stimuli once related to recurrent survival threats easier access for gaining control of the module through fear conditioning (Öhman & Mineka, 2001).

ISSUES FOR FURTHER RESEARCH

The claim that the fear module can be conditioned without awareness is a bold one given that there is a relative consensus in the field of human conditioning that awareness of the CS-US contingency is required for acquiring conditioned responses. However, as we have extensively argued elsewhere (Öhman & Mineka, 2001; Wiens & Öhman, 2002), there is good evidence that conditioning to nonconsciously presented CSs is possible if they are evolutionarily fear relevant. Other factors that might promote such nonconscious learning include intense USs, short CS-US intervals, and perhaps temporal overlap between the CS and the US. However, little research on these factors has been reported, and there is a pressing need to elaborate their relative effectiveness in promoting conditioning of the fear module outside of awareness.

One of the appeals of the fear module concept is that it is consistent with the current understanding of the neurobiology of fear conditioning, which gives a central role to the amygdala (e.g., Öhman & Mineka, 2001). However, this understanding is primarily based on animal data. Even though the emerging brain-imaging literature on human fear conditioning is consistent with this database, systematic efforts are needed in order to tie the fear module more convincingly to human brain mechanisms. For example, a conspicuous gap in knowledge concerns whether the amygdala is indeed specially tuned to conditioning contingencies involving evolutionarily fear-relevant CSs such as snakes.

An interesting question that can be addressed both at a psychological and at a neurobiological level concerns the perceptual mechanisms that give snake stimuli privileged access to the fear module. For example, are snakes detected at a lower perceptual threshold relative to non-fear-relevant objects? Are they identified faster than other objects once detected? Are they quicker to activate the fear module and attract attention once identified? Regardless of the locus of perceptual privilege, what visual features of snakes make them such powerful fear elicitors and attention captors? Because the visual processing in pathways preceding the cortical level is crude, the hypothesis that masked presentations of snakes directly access the amygdala implies that the effect is mediated by simple features of snakes rather than by the complex configuration of features defining a snake. Delineating these features would allow the construction of a "super fear stimulus." It could be argued that such a stimulus would depict "the archetypical evil" as represented in the human brain.

Recommended Reading

Mineka, S. (1992). Evolutionary memories, emotional processing, and the emotional disorders. *The Psychology of Learning and Motivation, 28*, 161–206.

Öhman, A., Dimberg, U., & Öst, L.-G. (1985). Animal and social phobias: Biological constraints on learned fear responses. In S. Reiss & R.R. Bootzin (Eds.), *Theoretical issues in behavior therapy* (pp. 123–178). New York: Academic Press.

Öhman; A., & Mineka, S. (2001). (See References)

Note

1. Address correspondence to Arne Öhman, Psychology Section, Department of Clinical Neuroscience, Karolinska Institute and Hospital, Z6:6, S-171 76 Stockholm, Sweden; e-mail: arne.ohman@cns.ki.se.

References

Agras, S., Sylvester, D., & Oliveau, D. (1969). The epidemiology of common fears and phobias. *Comprehensive Psychiatry, 10*, 151–156.

Cook, E.W., Hodes, R.L., & Lang, P.J. (1986). Preparedness and phobia: Effects of stimulus content on human visceral conditioning. *Journal of Abnormal Psychology, 95*, 195–207.

Cook, M., & Mineka, S. (1990). Selective associations in the observational conditioning of fear in rhesus monkeys. *Journal of Experimental Psychology: Animal Behavior Processes, 16*, 372–389.

Cook, M., & Mineka, S. (1991). Selective associations in the origins of phobic fears and their implications for behavior therapy. In P. Martin (Ed.), *Handbook of behavior therapy and psychological science: An integrative approach* (pp. 413–434). Oxford, England: Pergamon Press.

Kennedy, S.J., Rapee, R.M., & Mazurski, E.J. (1997). Covariation bias for phylogenetic versus ontogenetic fear-relevant stimuli. *Behaviour Research and Therapy, 35*, 415–422.

King, G.E. (1997, June). *The attentional basis for primate responses to snakes.* Paper presented at the annual meeting of the American Society of Primatologists, San Diego, CA.

Mineka, S., Keir, R., & Price, V. (1980). Fear of snakes in wild- and laboratory-reared rhesus monkeys (*Macaca mulatta*). *Animal Learning and Behavior, 8*, 653–663.

Öhman, A., Flykt, A., & Esteves, F. (2001). Emotion drives attention: Detecting the snake in the grass. *Journal of Experimental Psychology: General, 131*, 466–478.

Öhman, A., & Mineka, S. (2001). Fear, phobias and preparedness: Toward an evolved module of fear and fear learning. *Psychological Review, 108*, 483–522.

Öhman, A., & Soares, J.J.F. (1993). On the automatic nature of phobic fear: Conditioned electrodermal responses to masked fear-relevant stimuli. *Journal of Abnormal Psychology, 102*, 121–132.

Öhman, A., & Soares, J.J.F. (1994). "Unconscious anxiety": Phobic responses to masked stimuli. *Journal of Abnormal Psychology, 103*, 231–240.

Öhman, A., & Soares, J.J.F. (1998). Emotional conditioning to masked stimuli: Expectancies for aversive outcomes following nonrecognized fear-irrelevant stimuli. *Journal of Experimental Psychology: General, 127*, 69–82.

Sagan, C. (1977). *The dragons of Eden: Speculations on the evolution of human intelligence.* London: Hodder and Stoughton.

Tomarken, A.J., Sutton, S.K., & Mineka, S. (1995). Fear-relevant illusory correlations: What types of associations promote judgmental bias? *Journal of Abnormal Psychology, 104*, 312–326.

Wiens, S., & Öhman, A. (2002). Unawareness is more than a chance event: Comment on Lovibond and Shanks (2002). *Journal of Experimental Psychology: Animal Behavior Processes, 28*, 27–31.

Critical Thinking Questions

1. In many of the studies described here, subjects were shown the same pairings of snakes (or spiders) with shocks (or food) and flowers (or mushrooms) with shocks (or food). Subjects showed more fear, and indicated a higher correlation with shock, to the snakes. Do you think that comparing snakes/spiders to flowers/mushrooms represents a strong test of the authors' hypothesis? What about when they compare snakes to guns or to electrical equipment? What other kinds of things would you want to use in these experiments and why?
2. If the strong association between snakes and fear is the result of evolution, why don't all humans (or all monkeys) show it? If it goes back as far as the authors suggest, what else might you hypothesize that people should be predisposed to fear?
3. What role does consciousness play in evoking fear reactions to snakes?

Illusions of Control: How We Overestimate Our Personal Influence

Suzanne C. Thompson[1]
Department of Psychology, Pomona College, Claremont, California

Abstract

Illusions of control are common even in purely chance situations. They are particularly likely to occur in settings that are characterized by personal involvement, familiarity, foreknowledge of the desired outcome, and a focus on success. Person-based factors that affect illusions of control include depressive mood and need for control. One explanation of illusory control is that it is due to a control heuristic that is used to estimate control by assessing the factors of intentionality and connection to the outcome. Motivational influences on illusory control and consequences of overestimating one's control are also covered.

Keywords

illusions of control; perceived control

In an intriguing set of studies, Langer (1975) showed that people often overestimate their control even in situations governed purely by chance. In one of Langer's studies, some people were allowed to pick their own lottery ticket, and others got a ticket picked for them. Later, participants were given the option of exchanging their ticket for one in a lottery with more favorable odds. Despite the fact that exchanging the ticket increased the odds of winning, those who had selected their own lottery ticket number did not choose to exchange this personally chosen ticket. People seemed to think that choosing their own ticket increased their odds of winning as if their action of choosing their own ticket gave them some control over the outcome of the lottery. Similar *illusions of control* have been demonstrated in a variety of circumstances.

Alloy and Abramson (1979) extended the study of illusory control by manipulating factors that affect illusory control and measuring participants judgments of control. In these studies, people tried to get a green light to come on by pushing a button. They were told that the button might control the light, but, in actuality, there was no relationship between participants actions and whether the light came on. The light came on 25% of the time for participants in the low-reinforcement condition and 75% of the time for those in the high-reinforcement condition, regardless of how often or when the button was pushed. Estimates of personal control over the light were quite high in the high-reinforcement condition equivalent to a moderate to intermediate degree of control.

WHEN DO PEOPLE OVERESTIMATE CONTROL?

Do people always overestimate their control? Evidently not. You may have noted in the studies just described that the illusions of control occurred for people who selected their own lottery tickets or who were in the high-reinforcement condition. Participants in the other conditions were not so susceptible to thinking they had control. Both situational and person-based factors influence whether or not people will overestimate their control.

Situational factors include personal involvement, familiarity, foreknowledge of the desired outcome, and success at the task. *Personal involvement* refers to someone being the active agent as opposed to having others act for him or her. When people act for themselves, illusory control is likely, but when no personal action is involved for example, when someone else selects their tickets or throws the dice for them the sense of being able to control the outcome is greatly diminished. *Familiarity* is another important influence on illusions of control. When circumstances or the materials being worked with are familiar, it is easier to have inflated judgments of personal control than when the situation or task is new. *Foreknowledge of the desired outcome*, the extent to which people know the outcome they want when they act, is a third condition that affects whether or not personal control will be overestimated. In a prediction situation, actors know which outcome they want to achieve when they act. In contrast, in a postdiction situation, the action is taken before the actors know the outcome they would like. Illusions of control are stronger in prediction than in postdiction situations. *Success at the task* has a major influence on whether or not personal control is overestimated. Success versus failure has sometimes been manipulated experimentally by the number of times people get the outcome they are trying to achieve (e.g., the light comes on 75% vs. 25% of the time). Although the outcome is random and not in response to participants behavior, participants who get the outcome they want at a high rate have significantly and substantially higher estimates of their control than those who get the desired outcome at a low rate. People are also more likely to overestimate their control if the task involves acting to get a desired outcome than if the point is to avoid losing what they have. Evidently, the former type of task leads to a focus on successes and the latter highlights the losses. Even the patterning of successes and failures can affect the focus on success or failure. A string of early successes leads to higher estimates of personal control than does a string of early failures, even if the total number of successes over the entire session is the same.

The person factors that have an effect on illusory control include mood and need for control. In many circumstances, nondepressed individuals tend to overestimate their control, whereas those who are in a *depressed mood* have a more realistic assessment of their ability to control an outcome (Alloy & Abramson, 1979). These effects are found even when mood is manipulated experimentally. Nondepressed participants who are put in a depressed mood through a mood-induction technique make control judgments similar to those of participants who are chronically depressed (Alloy, Abramson, & Viscusi, 1981). Although the topic has not been studied extensively, there is some indication that a strong *need for control* can also lead people to overestimate their potential to affect outcomes. In an experimental demonstration of the effects of need for control, participants who were randomly assigned to a hungry condition were more likely than participants in a sated condition to think they could control a chance task for which a hamburger was a prize (Biner, Angle, Park, Mellinger, & Barber, 1995).

To sum up, people do not always overestimate their control, but when the situation is one that is familiar and focused on success, when the desired outcome is known and people are taking action for themselves, the setup is ripe for illusory control. In addition, nondepressed individuals and those with a need for the outcome are particularly prone to overestimate their control.

WHY DO PEOPLE OVERESTIMATE THEIR CONTROL?

The question of why illusions of control occur was first addressed by Langer (1975), who proposed that skill and chance situations are easily confused. This confusion is especially likely to happen when chance situations have the trappings that are characteristic of skill-based situations (e.g., familiarity, choice, involvement). However, there are several reasons why a confusion of skill and chance situations is not by itself a good explanation for the variety of conditions under which illusions of control appear. For example, the factors that increase illusory control are broader than ones that reflect a confusion between chance and controllable situations. Illusory control is also more likely in situations that involve a focus on success rather than failure, as well as among individuals who are in a positive rather than depressed mood and who have a need for control. Also, feedback that highlights failures and negative moods can eliminate or reverse illusory judgments of control. None of these effects is easily explained by a confusion of skill and chance elements.

My colleagues and I have offered a more complete explanation of why illusions of control occur (Thompson, Armstrong, & Thomas, 1998). We propose that people use a control heuristic to judge their degree of influence over an outcome. A heuristic is a shortcut or simple rule that can be used to reach a judgment, in this case, an estimate of one's control over achieving an outcome. The control heuristic involves two elements: one's intention to achieve the outcome and the perceived connection between one's action and the desired outcome. Both intention and connection are cues that are used to judge how much control one has. If one intended an outcome and can see a connection between one's own action and the outcome, then perceptions of personal control are high.

Like most heuristics, this simple rule will often lead to accurate judgments because intentionality and a perceived connection (e.g., seeing that the outcome immediately follows one's action) often occur in situations in which one does, in fact, have control. However, the heuristic can also lead to overestimations of control because intentionality and connection can occur in situations in which a person has little or no control. For example, gamblers playing the slot machines in Las Vegas pull the handles with the intention of getting a winning combination to slide into place. When their handle pulling is followed by the desired outcome, a connection is established between their action and the outcome. Because intentionality and connection are strong, the stage is set for gamblers to think that they have some control in this situation.

The control heuristic provides a unifying explanation for all of the various factors that have been found to affect illusions of control. Each is influential because it affects perceived intentionality or the connection between one's behavior and the outcome. For example, personal involvement is essential for the illusion of control because connection between an actor's actions and an event cannot be observed or imagined unless the actor acts. Foreknowledge is important for illusory control because without it the actor cannot intend to effect a particular outcome. Success-focused tasks increase illusions of control because these circumstances lead actors to overestimate the connection between their action and the successful outcome. They do this because compared with tasks focusing on failure, success-oriented tasks produce more instances in which actor's actions are followed by the desired outcome and because these types of tasks direct attention to success; both of these factors increase perceptions of the number of positive confirming cases. Failure experiences and a focus on losing have the opposite effect by highlighting the times when actors' behavior is not followed by the desired outcome, thereby weakening the perception of connection. Depressed mood may be associated with lower assessments of personal control because people who are depressed focus on failure, not on success. Because of this focus on failure, they are less likely than nondepressed people to notice and overestimate the connection between their own behavior and successes.

DOES WANTING TO HAVE CONTROL LEAD TO ILLUSORY CONTROL?

Because the benefits of believing oneself to have control (e.g., positive mood and increased motivation) may be realized even if one's control is illusory, it seems reasonable to suggest that people are often motivated to overestimate their control. Some recent studies in my lab have examined the effects of a motive for control on illusory control judgments. Participants worked on a computer task that was similar to the light task described earlier. Motive for control was manipulated by paying some participants for each time a green computer screen appeared. Despite the fact that participants had no control, the motive for control had a strong effect on control judgments: Participants who were paid for each appearance of the green screen had significantly higher perceptions of control than participants lacking this motive.

According to the control-heuristic model, motives for control affect judgments of control by biasing either judgments of intentionality or judgments of connection. For example, believing that one could have foreseen the outcome heightens a sense of acting intentionally. Therefore, for individuals who are motivated to have control, judgments of intentionality can be heightened through the hindsight bias, in which individuals overestimate the degree to which they could have anticipated an outcome. Counterfactuals, or imagined alternative outcomes to an event, are another route for a control motive to increase illusions of control. Perceptions of connection can be strengthened through the use of controllable counterfactuals because imagining undoing antecedents that are controllable by the actor heightens the connection between action and outcome. For example, a lottery winner who thinks, "If I hadn't thought to buy that ticket, I wouldn't have won $1,000" highlights the connection between her action and the desired outcome, whereas imagining a counterfactual with an uncontrollable element would lessen the connection.

THE CONSEQUENCES OF ILLUSORY CONTROL

Not all people overestimate their personal control. Moderately depressed individuals tend to have a realistic sense of how much they are contributing to an outcome. Does that mean that people are better off if they overestimate their personal control?

Only a few studies have addressed the issue of the adaptiveness of illusory control per se. Some of these studies have found that illusory control enhances adaptive functioning. A study by Alloy and Clements (1992) examined college students who varied in the degree to which they exhibited the illusion of control. Compared with students who tended not to have illusions of control, students who displayed greater illusions of control had less negative mood after a failure on a lab task, were less likely to become discouraged when they subsequently experienced negative life events, and were less likely to get depressed a month later, given the occurrence of a high number of negative life events. Thus, individuals who are susceptible to an illusion of control may be at a decreased risk for depression and discouragement in comparison to those individuals who are not.

In contrast to this positive finding, there is also evidence that overestimating one's control is not adaptive. For example, Donovan, Leavitt, and Walsh (1990) investigated the influence of illusory control on child-care outcomes. The degree of illusory control was measured with a simulated child-care task in which mothers tried to terminate the crying of an audio-taped baby. Mothers judged their control over stopping the tape by pushing a button. In actuality, their responses were not connected to the operation of the tape. A subsequent simulation assessed the mothers ability to learn effective responses for getting an infant to stop crying. Mothers with a high illusion of control on the first task were more susceptible to helplessness on the second task, for example, by not responding even when control was possible.

Which is the correct view: that illusory thinking is generally useful because it leads to positive emotions and motivates people to try challenging tasks, or that people are better off if they have an accurate assessment of themselves and their situation? Perhaps a third possibility is correct: Sometimes illusory control is adaptive, and at other times it is not. For example, illusions of control may be reassuring in stressful situations, but lead people to take unnecessary risks when they occur in a gambling context. The challenge for researchers is to examine the consequences of illusory control in a variety of situations to answer this important question.

Recommended Reading

Alloy, L.B., & Abramson, L.Y. (1979). (See References)
Langer, E.J. (1975). (See References)
Thompson, S.C., Armstrong, W., & Thomas, C. (1998). (See References)

Note

1. Address correspondence to Suzanne Thompson, Department of Psychology, Pomona College, Claremont, CA 91711.

References

Alloy, L.B., & Abramson, L.Y. (1979). Judgment of contingency in depressed and nondepressed students: Sadder but wiser? *Journal of Experimental Psychology: General*, *108*, 441–485.

Alloy L.B., Abramson, L.Y., & Viscusi, D. (1981). Induced mood and the illusion of control. *Journal of Personality and Social Psychology*, *41*, 1129–1140.

Alloy, L.B., & Clements, C.M. (1992). Illusion of control: Invulnerability to negative affect and depressive symptoms after laboratory and natural stressors. *Journal of Abnormal Psychology*, *307*, 234–245.

Biner, P.M., Angle, S.T., Park, J.H., Mellinger, A.E., & Barber, B.C. (1995). Need and the illusion of control. *Personality and Social Psychology Bulletin*, *23*, 899–907.

Donovan, W.L., Leavitt, L.A., & Walsh, R.O. (1990). Maternal self-efficacy: Illusory control and its effect on susceptibility to learned helplessness. *Child Development*, *61*, 1638–1647.

Langer, E.J. (1975). The illusion of control. *Journal of Personality and Social Psychology*, *32*, 311–328.

Thompson, S.C., Armstrong, W., & Thomas, C. (1998). Illusions of control, underestimations, and accuracy: A control heuristic explanation. *Psychological Bulletin*, *123*, 143–161.

Critical Thinking Questions

1. What kinds of people are more or less likely to show the illusion of control? If depressed people believe that they have less control, how do we know that the depression is causing the belief and not that the belief is (at least in part) causing the depression?

2. Besides deciding to become depressed, what could you do to decrease the illusion of control and increase your accuracy when you have to make these kinds of judgments?

3. How would you currently answer the question: Is illusory thinking adaptive? What else would you want to know before you would be happy with your answer? Do you think that the question is ultimately necessarily answerable?

When Possibility Informs Reality: Counterfactual Thinking as a Cue to Causality

Barbara A. Spellman[1] and David R. Mandel
Department of Psychology, University of Virginia, Charlottesville, Virginia (B.A.S.), and Department of Psychology, University of Hertfordshire, Hatfield, Hertfordshire, England (D.R.M.)

Abstract

People often engage in counterfactual thinking, that is, imagining alternatives to the real world and mentally playing out the consequences. Yet the counterfactuals people tend to imagine are a small subset of those that could possibly be imagined. There is some debate as to the relation between counterfactual thinking and causal beliefs. Some researchers argue that counterfactual thinking is the key to causal judgments; current research suggests, however, that the relation is rather complex. When people think about counterfactuals, they focus on ways to prevent bad or uncommon outcomes; when people think about causes, they focus on things that covary with outcomes. Counterfactual thinking may affect causality judgments by changing beliefs about the probabilities of possible alternatives to what actually happened, thereby changing beliefs as to whether a cause and effect actually covary. The way in which counterfactual thinking affects causal attributions may have practical consequences for mental health and the legal system.

Keywords

counterfactuals; causality; reasoning

> If I *were* in New York right now, and if I happened to be standing on a sidewalk on my way to lunch, waiting for the light to change, and if a car happened to jump the curb, I might be struck dead. By not being there, I may have freed that space on the sidewalk for someone else who might be standing there and get run over. And in that event you might say that I'm partially responsible for that death. In a weak sense I'd be responsible at the end of a long causal chain. We're all linked by these causal chains to everyone around us. . . .
>
> McInerney (1992, p. 120)

Most of us have outlandish suppositional "what if" thoughts like this from time to time. We may wonder, "What if my parents had never met?" (*Back to the Future*) or "What if I had never been born?" (*It's a Wonderful Life*). Seemingly less outlandish and far more ubiquitous are our thoughts about the ways in which things might have been different "if only" an event leading up to a particular outcome had been different. For instance, suppose that you get into a car accident while taking a scenic route home that you rarely use but chose to take today because it was particularly sunny. If you are like many other people, you might "undo" the accident by thinking something like "If only I had taken my usual route, I wouldn't have been in the accident" (Kahneman & Tversky, 1982; Mandel & Lehman, 1996). These "if only" thoughts are termed *counterfactual conditionals* because they focus on a changed (or *mutated*) version of reality (i.e., one that is counter to fact) and because they represent a conditional if-then relation between an antecedent event (e.g., the route taken) and a consequent event (e.g., getting into an accident).

CAUSAL CONSEQUENCES OF COUNTERFACTUAL THINKING

Counterfactual thinking has consequences beyond mere daydreams or entertainment. Engaging in counterfactual thinking may influence people's causal attributions.

Practical consequences of this relation can be seen for mental health and the legal system.

For example, counterfactual thinking can produce what Kahneman and Miller (1986) termed *emotional amplification*—a heightening of affect brought about by realizing that an outcome was not inevitable because it easily could have been undone. Counterfactual thinking can amplify feelings of regret, distress and self-blame, and shame and guilt, as well as satisfaction and happiness (see Roese & Olson, 1995, for reviews). Perhaps recurrent counterfactual thinking, in which people obsessively mutate their own actions, leads people to exaggerate their own causal role for both their misfortunes and their successes, resulting in those heightened emotions.

Counterfactual thinking can also affect liability and guilt judgments in legal settings. For example, in one study, subjects read a story about a date rape and then listened to a mock lawyer's closing argument suggesting possible mutations to the story (Branscombe, Owen, Garstka, & Coleman, 1996). If the argument mutated the defendant's actions so that the rape would be undone, the rapist was assigned more fault (cause, blame, and responsibility) than if his actions were mutated but the rape still would have occurred. Similarly, when the victim's actions were mutated, she was assigned more fault if the mutation would undo the rape than if the rape still would have occurred.

THE METHOD BEHIND OUR MUSINGS

Although in our counterfactual musings there are an infinite number of ways we can mutate antecedents so that an outcome would be different, we tend to introduce relatively minor mutations that are systematically constrained (see Roese, 1997). People commonly mutate abnormal events by restoring them to their normal default (Kahneman & Miller, 1986). For example, in the car-accident story, people undo the accident by restoring the abnormal antecedent of taking the unusual route to its normal default—taking the usual route (Kahneman & Tversky, 1982). People also tend to mutate antecedents that are personally controllable (see Roese, 1997). For instance, one is more likely to think, "If only I had decided to take the usual route . . ." than to think, "If only it wasn't sunny" or even "If only the other driver was not reckless" (Mandel & Lehman, 1996). More generally, research on counterfactual thinking demonstrates that people entertain a very small and nonrandom sample from the universe of possible mutations.

Just as various antecedent events leading up to an outcome are differentially likely to be mutated, various outcomes are differentially likely to trigger counterfactual thinking. Negative events and abnormal events are the most likely triggers (see Roese, 1997; Roese & Olson, 1995). Consequently, research on counterfactual thinking has focused on subjects' counterfactual thoughts about negative events like car accidents, food poisonings, and stock-market losses, and unexpected positive events like winning a lottery.

COUNTERFACTUAL THINKING AS TESTS OF CAUSALITY

Kahneman and Tversky's (1982) highly influential chapter on the *simulation heuristic* described the basic logic by which if-only thinking could be used to explore the causes of past events. They proposed that people often run an if-then simulation in their minds to make various sorts of judgments, including predictions of future events and causal explanations of past events. Accordingly, to understand whether X caused Y, a person may imagine that X had not occurred and then imagine what events would follow. The easier it is to imagine that Y would not now follow, the more likely the person is to view X as a cause of Y.

Kahneman and Tversky's (1982) proposal is certainly plausible: They were not suggesting that all counterfactual-conditional thinking is necessarily directed at understanding the causes of past events, but simply making the point that such thinking could be useful toward that end. Nor did they argue that all forms of causal reasoning are equally likely to rely on counterfactual thinking; their examples suggested that counterfactual thinking would be most useful for understanding the causes of a single case (as in law) rather than for understanding causal relations governing a set of events (as in science). Indeed, their proposal echoed earlier work on causal reasoning by the philosopher Mackie (1974) and the legal philosophers Hart and Honoré (1959/1985), and amounted to what in political science and law is assumed generally to be an important criterion for attributing causality: In order to ascribe causal status to an actor or event, one should believe that *but for* the causal candidate in question, the outcome would not have occurred. In other words, one must demonstrate that the causal candidate was *necessary*, under the circumstances, for the outcome to occur. Despite the modest scope of Kahneman and Tversky's proposal, however, many attribution researchers subsequently posited that counterfactual thinking plays a primary role in causal attributions.

EMPIRICAL EVIDENCE

The first empirical study to measure both counterfactual thinking and judgments of causality in the same story (Wells & Gavanski, 1989) supported the idea that mutability was the key to causality. Subjects read about a woman who was taken to dinner by her boss. The boss ordered for both of them, but the dish he ordered contained wine—to which the woman was allergic. She ate the food and died. The boss had considered ordering something else: In one version, the alternative dish also contained wine; in the other version, the alternative dish did not contain wine. Subjects were asked to list four things that could have been different in the story to prevent the woman's death (mutation task) and to rate how much of a causal role the boss's ordering decision played in her death. Relative to subjects who read that both dishes contained wine, subjects who read that the other dish did not contain wine (a) were more likely to mutate the boss's decision and (b) rated the boss's decision as more causal. These results were interpreted as showing two things: that for causality to be assigned to an event (the boss's decision), the event must have a counterfactual that would have undone the outcome (i.e., mutability is necessary for causality) and that the more available a counterfactual alternative is, the more causal that event seems (i.e., ease of mutability affects perceived causal impact).

Later research, however, has found different patterns for mutability and causality in other, richer, scenarios. For example, in another study (Mandel & Lehman, 1996), subjects read a story about a car accident like the one alluded to earlier (based on Kahneman & Tversky, 1982): On a sunny day, Mr. Jones decided to take an unusual route home; being indecisive, he braked hard to stop at a yellow light; he began driving when the light turned green; he was hit by a drunk teenager who charged through the red light. One group of subjects was asked how Jones would finish the thought "If only . . . ," and another group was asked what Jones would think caused the accident. The subjects tended to mutate Jones's decision to take the unusual route and his indecisive driving but to assign causality to the teenager's drunk driving. Hence, the most mutated event need not be perceived as the most causal event.

Therefore, although sometimes mutability and causality converge (as in the food decision), sometimes they differ (as in the car accident). Yet the latter experiment does not disconfirm that mutability is necessary for causality: The drunk driver's actions—the most causal event—can easily be mutated to undo the outcome (even though other aspects of the story are mutated more often).

However, there is a situation that demonstrates that mutability is *not necessary* for causality. Suppose that Able and Baker shoot Smith at the exact same instant. Smith dies and the coroner reports that either shot alone would have killed him (i.e., there is *causal overdetermination*). Able can argue that he did not kill Smith because even if he had not shot Smith, Smith would have died anyway. Mutating Able's actions alone does not work; in fact, subjects mutate Able's and Baker's actions together to undo Smith's death. But when attributing causality, subjects judge Able and Baker individually to be causes of Smith's death and sentence Able and Baker each to the maximum jail time (Spellman, Maris, & Wynn, 1999). Both philosophers and legal theorists consider such cases of causal overdetermination to be an exception to the but-for causality rule. Because causality is assigned even when mutation does not undo the outcome, mutability is not necessary for causality.

Moreover, mutability is *not sufficient* for causality. For example, one might believe that Jones's car accident would not have occurred if it had not been a sunny day (as Jones would have taken his normal route); however, no one would argue that the sunny day caused the accident. Obviously, therefore, not everything that can be mutated to undo an outcome is judged as a cause of that outcome.

RECENT THEORETICAL PERSPECTIVES

The relation between mutability and causality is not simply one of identity, necessity, or sufficiency; rather, a more complex picture has emerged.

According to a recent proposal (Mandel & Lehman, 1996), everyday counterfactuals are not tests of causality. Rather, for negative outcomes, counterfactuals are after-the-fact realizations of ways that would have been sufficient to prevent the outcomes—and especially ways that actors themselves could have prevented their misfortunes. Data from the car-accident experiment reported earlier support this account. In addition to the counterfactual- and causal-question groups, a third group of subjects was included in this experiment. These subjects answered a question about how Jones would think the accident could have been prevented. Not surprisingly, subjects' answers for the preventability question focused on the same aspects of the story as the answers for the counterfactual question, but were different from the answers for the causality question. Thus, although, logically, sufficient-but-foregone preventors (e.g., taking the usual route) seem like necessary causes, psychologically, attributions of preventability and causality are different. In attributing preventability, people focus on controllable antecedents (e.g., choice of route, stopping at a yellow light); in attributing causality, people focus on antecedents that general knowledge suggests would covary with, and therefore predict, the outcome (e.g., drunk drivers).

The *crediting causality* model (Spellman, 1997) tells us how a covariation analysis of causal candidates might be performed in reasoning about single cases. According to this model, causality attributions in single cases are analogous to causality attributions in science. In science, to find whether something is causal (e.g., whether drunk driving causes accidents), people presumably compare the probability of the effect when the cause is present (e.g., the probability of accidents given drunk drivers) with the probability of the effect when the cause is absent (e.g., the probability of accidents given no drunk drivers); if the former probability is larger, people infer causality. For the single case, people may consider and compare that probability with the probability of an accident before that event occurred (i.e., the baseline probability). If the event (e.g., a drunk driver) raises the probability of the outcome, then it is causal; if it does not (e.g., a sunny day), it is not.

Counterfactual thinking might affect causal attributions by acting as input to that probability comparison. If one can counterfactually imagine many alternatives to an outcome like an accident, then the baseline probability of the accident seems low, and, therefore, the causality of the event in question (e.g., drunk driver) seems high. If there are no alternatives to an outcome (e.g., all foods contain wine and will cause illness), then the baseline probability of the outcome is high, and the causality of the event in question (e.g., ordering a particular dish) seems low. Similarly, in the experiment involving counterfactual thinking about the date rape (Branscombe et al., 1996), thinking about mutations that would undo the rape may have changed estimates of the baseline probability of a rape, and thus affected the subjects' legal judgments. Note that when it seems that nothing can be mutated to undo the outcome, people feel the outcome was inevitable; that is, it was due to fate (Mandel & Lehman, 1996).

CONCLUDING REMARKS

The philosopher David Hume described causation—or, more accurately, causal thinking—as the "cement of the universe" because it binds together people's perceptions of events that would otherwise appear unrelated. It seems, however, that our beliefs about the universe of the actual, to which Hume referred, are affected by our considerations of the merely possible—created by the "what ifs" and "if onlys" of counterfactual thinking.

Recommended Reading

Kahneman, D., & Miller, D.T. (1986). (See References)
Mackie, J.L. (1974). (See References)
Mandel, D.R., & Lehman, D.R. (1996). (See References)
Roese, N.J. (1999). *Counterfactual Research News* [On-line]. Available: http://www.psych.nwu.edu/psych/people/faculty/roese/research/cf/cfnews.htm
Roese, N.J., & Olson, J.M. (Eds.). (1995). (See References)

Acknowledgments—This work was supported in part by a NIMH FIRST Award to the first author and by Research Grant No. 95-3054 from the Natural Sciences and Engineering Research Council of Canada to the second author. We thank Darrin Lehman and Angeline Lillard for helpful comments on an earlier draft.

Note

1. Address correspondence to Barbara A. Spellman, Department of Psychology, 102 Gilmer Hall, University of Virginia, Charlottesville, VA 22903; email: spellman@virginia.edu.

References

Branscombe, N.R., Owen, S., Garstka, T.A., & Coleman, J. (1996). Rape and accident counterfactuals: Who might have done otherwise and would it have changed the outcome? *Journal of Applied Social Psychology*, *26*, 1042–1067.
Hart, H.L., & Honoré, A.M. (1985). *Causation in the law* (2nd ed.). Oxford, England: Oxford University Press. (Original work published 1959)
Kahneman, D., & Miller, D.T. (1986). Norm theory: Comparing reality to its alternatives. *Psychological Review*, *93*, 136–153.
Kahneman, D., & Tversky, A. (1982). The simulation heuristic. In D. Kahneman, P. Slovic, & A. Tversky (Eds.), *Judgment under uncertainty: Heuristics and biases* (pp. 201–208). New York: Cambridge University Press.
Mackie, J.L. (1974). *The cement of the universe: A study of causation*. Oxford, England: Oxford University Press.
Mandel, D.R., & Lehman, D.R. (1996). Counterfactual thinking and ascriptions of cause and preventability. *Journal of Personality and Social Psychology*, *71*, 450–463.
McInerney, J. (1992). *Brightness falls*. New York: Knopf.
Roese, N.J. (1997). Counterfactual thinking. *Psychological Bulletin*, *121*, 133–148.
Roese, N.J., & Olson, J.M. (Eds.). (1995). *What might have been: The social psychology of counterfactual thinking*. Mahwah, NJ: Erlbaum.
Spellman, B.A. (1997). Crediting causality. *Journal of Experimental Psychology: General*, *126*, 323–348.
Spellman, B.A., Maris, J.R., & Wynn, J.E. (1999). *Causality without mutability*. Unpublished manuscript, University of Virginia, Charlottesville.
Wells, G.L., & Gavanski, I. (1989). Mental simulation of causality. *Journal of Personality and Social Psychology*, *56*, 161–169.

Critical Thinking Questions

1. Imagine that you are an attorney defending your client, the X cigarette company, against a law suit by a plaintiff who has smoked X cigarettes for years and has recently developed lung cancer. The plaintiff is also suing companies Y and Z because he smoked their cigarettes, too. Proving causality is essential in this kind of law suit. Think of several ways in which you might defend your client against the claim that "Cigarette X caused the plaintiff's lung cancer". Do you always seem to get to the right or just result?
2. This article claims that "there are an infinite number of ways we can mutate antecedents so that an outcome would be different." Consider the car accident story in the article. Can you think of an infinite number of things to mutate? (Hint: get out of the box.) Why do you think people tend to converge on the same set of minor mutations (e.g., taking the usual rather than unusual route) and do not often consider the lack of Martian intervention?
3. Suppose you convince your friend to skip class and go skiing with you. Your friend hits a patch of ice, the binding does not release, and she breaks her leg. What counterfactuals come to mind? Which make you feel regret? What causes come to mind? Do the lists of counterfactuals and causes overlap? Which do, which don't, and why?

Culture and Causal Cognition

Ara Norenzayan and Richard E. Nisbett[1]
Centre de Récherche en Epistemologie Appliquée, Ecole Polytechnique, Paris, France (A.N.), and Department of Psychology, University of Michigan, Ann Arbor, Michigan (R.E.N.)

Abstract

East Asian and American causal reasoning differs significantly. East Asians understand behavior in terms of complex interactions between dispositions of the person or other object and contextual factors, whereas Americans often view social behavior primarily as the direct unfolding of dispositions. These culturally differing causal theories seem to be rooted in more pervasive, culture-specific mentalities in East Asia and the West. The Western mentality is analytic, focusing attention on the object, categorizing it by reference to its attributes, and ascribing causality based on rules about it. The East Asian mentality is holistic, focusing attention on the field in which the object is located and ascribing causality by reference to the relationship between the object and the field.

Keywords

causal attribution; culture; attention; reasoning

Psychologists within the cognitive science tradition have long believed that fundamental reasoning processes such as causal attribution are the same in all cultures (Gardner, 1985). Although recognizing that the content of causal beliefs can differ widely across cultures, psychologists have assumed that the ways in which people come to make their causal judgments are essentially the same, and therefore that they tend to make the same sorts of inferential errors. A case in point is the fundamental attribution error, or FAE (Ross, 1977), a phenomenon that is of central importance to social psychology and until recently was held to be invariable across cultures.

The FAE refers to people's inclination to see behavior as the result of dispositions corresponding to the apparent nature of the behavior. This tendency often results in error when there are obvious situational constraints that leave little or no role for dispositions in producing the behavior. The classic example of the FAE was demonstrated in a study by Jones and Harris (1967) in which participants read a speech or essay that a target person had allegedly been required to produce by a debate coach or psychology experimenter. The speech or essay favored a particular position on an issue, for example, the legalization of marijuana. Participants' estimates of the target's actual views on the issue reflected to a substantial extent the views expressed in the speech or essay, even when they knew that the target had been explicitly instructed to defend a particular position. Thus, participants inferred an attitude that corresponded to the target person's apparent behavior, without taking into account the situational constraints operating on the behavior. Since that classic study, the FAE has been found in myriad studies in innumerable experimental and naturalistic contexts, and it has been a major focus of theorizing and a continuing source of instructive pedagogy for psychology students.

CULTURE AND THE FAE

It turns out, however, that the FAE is much harder to demonstrate with Asian populations than with European-American populations (Choi, Nisbett, & Norenzayan, 1999). Miller (1984) showed that Hindu Indians preferred to explain ordinary life events in terms of the situational context in which they occurred, whereas Americans were much more inclined to explain similar events in terms of presumed dispositions. Morris and Peng (1994) found that Chinese newspapers and Chinese students living in the United States tended to explain murders (by both Chinese and American perpetrators) in terms of the situation and even the societal context confronting the murderers, whereas American newspapers and American students were more likely to explain the murders in terms of presumed dispositions of the perpetrators.

Recently Jones and Harris's (1967) experiment was repeated with Korean and American participants (Choi et al., 1999). Like Americans, the Koreans tended to assume that the target person held the position he was advocating. But the two groups responded quite differently if they were placed in the same situation themselves before they made judgments about the target. When observers were required to write an essay, using four arguments specified by the experimenter, the Americans were unaffected, but the Koreans were greatly affected. That is, the Americans' judgments about the target's attitudes were just as much influenced by the target's essay as if they themselves had never experienced the constraints inherent in the situation, whereas the Koreans almost never inferred that the target person had the attitude expressed in the essay.

This is not to say that Asians do not use dispositions in causal analysis or are not occasionally susceptible to the FAE. Growing evidence indicates that when situational cues are not salient, Asians rely on dispositions or manifest the FAE to the same extent as Westerners (Choi et al., 1999; Norenzayan, Choi, & Nisbett, 1999). The cultural difference seems to originate primarily from a stronger East Asian tendency to recognize the causal power of situations.

The cultural differences in the FAE seem to be supported by different folk theories about the causes of human behavior. In one study (Norenzayan et al., 1999), we asked participants how much they agreed with paragraph descriptions of three different philosophies about why people behave as they do: (a) a strongly dispositionist philosophy holding that "how people behave is mostly determined by their personality," (b) a strongly situationist view holding that behavior "is mostly determined by the situation" in which people find themselves, and (c) an interactionist view holding that behavior "is always jointly determined by personality and the situation." Korean and American participants endorsed the first position to the same degree, but Koreans endorsed the situationist and interactionist views more strongly than did Americans.

These causal theories are consistent with cultural conceptions of personality as well. In the same study (Norenzayan et al., 1999), we administered a scale designed to measure agreement with two different theories of personality: entity theory, or the belief that behavior is due to relatively fixed dispositions such as traits, intelligence, and moral character, and incremental theory, or the belief that behavior is conditioned on the situation and that any relevant dispositions are subject to change (Dweck, Hong, & Chiu, 1993). Koreans for the most part rejected entity theory, whereas Americans were equally likely to endorse entity theory and incremental theory.

ANALYTIC VERSUS HOLISTIC COGNITION

The cultural differences in causal cognition go beyond interpretations of human behavior. Morris and Peng (1994) showed cartoons of an individual fish moving in a variety of configurations in relation to a group of fish and asked participants why they thought the actions had occurred. Chinese participants were inclined to attribute the behavior of the individual fish to factors external to the fish (i.e., the group), whereas American participants were more inclined to attribute the behavior of the fish to internal factors. In studies by Peng and Nisbett (reported in Nisbett, Peng, Choi, & Norenzayan, in press), Chinese participants were shown to interpret even the behavior of schematically drawn, ambiguous physical events—such as a round object dropping through a surface and returning to the surface—as being due to the relation between the object and the presumed medium (e.g., water), whereas Americans tended to interpret the behavior as being due to the properties of the object alone.

The Intellectual Histories of East Asia and Europe

Why should Asians and Americans perceive causality so differently? Scholars in many fields, including ethnography, history, and philosophy of science, hold that, at least since the 6th century B.C., there has been a very different intellectual tradition in the West than in the East (especially China and those cultures, like the Korean and Japanese, that were heavily influenced by China; Nisbett et al., in press). The ancient Greeks had an *analytic* stance: The focus was on categorizing the object with reference to its attributes and explaining its behavior using rules about its category memberships. The ancient Chinese had a holistic stance, meaning that there was an orientation toward the field in which the object was found and a tendency to explain the behavior of the object in terms of its relations with the field.

In support of these propositions, there is substantial evidence that early Greek and Chinese science and mathematics were quite different in their strengths and weaknesses. Greek science looked for universal rules to explain events and was concerned with categorizing objects with respect to their essences. Chinese science (some people would say it was a technology only, though a technology vastly superior to that of the Greeks) was more pragmatic and concrete and was not concerned with foundations or universal laws. The difference between the Greek and Chinese orientations is well captured by Aristotle's physics, which explained the behavior of an object without reference to the field in which it occurs. Thus, a stone sinks into water because it has the property of gravity, and a piece of wood floats because it has the property of levity. In contrast, the principle that events always occur in some context or field of forces was understood early on in China.

Some writers have suggested that the mentality of East Asians remains more holistic than that of Westerners (e.g., Nakamura, 1960/1988). Thus, modern East Asian laypeople, like the ancient Chinese intelligentsia, are attuned to the field and the overall context in determining events. Western civilization was profoundly shaped by ancient Greece, so one would expect the Greek intellectual stance of object focus to be widespread in the West.

Attention to the Field Versus the Object

If East Asians tend to believe that causality lies in the field, they would be expected to attend to the field. If Westerners are more inclined to believe that causality inheres in the object, they might be expected to pay relatively more attention to the object than to the field. There is substantial evidence that this is the case.

Attention to the field as a whole on the part of East Asians suggests that they might find it relatively difficult to separate the object from the field. This notion rests on the concept of *field dependence* (Witkin, Dyk, Faterson, Goodenough, & Karp, 1974). Field dependence refers to a relative difficulty in separating objects from the context in which they are located. One way of measuring field dependence is by means of the rod-and-frame test. In this test, participants look into a long rectangular box at the end of which is a rod. The rod and the box frame can be rotated independently of one another, and participants are asked to state when the rod is vertical. Field dependence is indicated by the extent to which the orientation of the frame influences judgments of the verticality of the rod. The judgments of East Asian (mostly Chinese) participants have been shown to be more field dependent than those of American participants (Ji, Peng, & Nisbett, in press).

In a direct test of whether East Asians pay more attention to the field than Westerners do (Masuda & Nisbett, 1999), Japanese and American participants saw underwater scenes that included one or more *focal* fish (i.e., fish that were larger and faster moving than other objects in the scene) among many other objects, including smaller fish, small animals, plants, rocks, and coral. When asked to recall what they had just viewed, the Japanese and American participants reported equivalent amounts of detail about the focal fish, but the Japanese reported far more detail about almost everything else in the background and made many more references to interactions between focal fish and background objects. After watching the scenes, the participants were shown a focal fish either on the original background or on a new one. The ability of the Japanese to recognize a particular focal fish was impaired if the fish was shown on the "wrong" background. Americans' recognition was uninfluenced by this manipulation.

ORIGINS OF THE CULTURAL DIFFERENCE IN CAUSAL COGNITION

Most of the cross-cultural comparisons we have reviewed compared participants who were highly similar with respect to key demographic variables, namely, age, gender, socioeconomic status, and educational level. Differences in cognitive abilities were controlled for or ruled out as potential explanations for the data in studies involving a task (e.g., the rod-and-frame test) that might be affected by such abilities. Moreover, the predicted differences emerged regardless of whether the East Asians were tested in their native languages in East Asian countries or tested in English in the United States. Thus, the lack of obvious alternative explanations, combined with positive evidence from intellectual history and the convergence of the data across a diverse set of studies (conducted in laboratory as well as naturalistic contexts), points to culturally shared causal theories as the most likely explanation for the group differences.

But why might ancient societies have differed in the causal theories they produced and passed down to their contemporary successor cultures? Attempts to answer such questions must, of course, be highly speculative because they involve complex historical and sociological issues. Elsewhere, we have summarized the views of scholars who have suggested that fundamental differences between societies may result from ecological and economic factors (Nisbett et al., in press). In China, people engaged in intensive farming many centuries before Europeans did. Farmers need to be cooperative with one another, and their societies tend to be collectivist in nature. A focus on the social field may generalize to a holistic understanding of the world. Greece is a land where the mountains descend to the sea and large-scale agriculture is not possible. People earned a living by keeping animals, fishing, and trading. These occupations do not require so much intensive cooperation, and the Greeks were in fact highly individualistic. Individualism in turn encourages attending only to the object and one's goals with regard to it. The social field can be ignored with relative impunity, and causal perception can focus, often mistakenly, solely on the object. We speculate that contemporary societies continue to display these mentalities because the social psychological factors that gave rise to them persist to this day.

Several findings by Witkin and his colleagues (e.g., Witkin et al., 1974), at different levels of analysis, support this historical argument that holistic and analytic cognition originated in collectivist and individualist orientations, respectively. Contemporary farmers are more field dependent than hunters and industrialized peoples; American ethnic groups that operate under tighter social constraints are more field dependent than other groups; and individuals who are attuned to social relationships are more field dependent than those who are less focused on social relationships.

FUTURE DIRECTIONS

A number of questions seem particularly interesting for further inquiry. Should educational practices take into account the differing attentional foci and causal theories of members of different cultural groups? Can the cognitive skills characteristic of one cultural group be transferred to another group? To what extent can economic changes transform the sort of cultural-cognitive system we have described? These and other questions about causal cognition will provide fertile ground for research in the years to come.

Recommended Reading

Choi, I., Nisbett, R.E., & Norenzayan, A. (1999). (See References)

Fiske, A., Kitayama, S., Markus, H.R., & Nisbett, R.E. (1998). The cultural matrix of social psychology. In D.T. Gilbert, S.T. Fiske, & G. Lindzey (Eds.), The handbook of social psychology (4th ed., Vol. 2, pp. 915–981). Boston: McGraw-Hill.

Lloyd, G.E.R. (1996). Science in antiquity: The Greek and Chinese cases and their relevance to problems of culture and cognition. In D.R. Olson & N. Torrance (Eds.), *Modes of thought: Explorations in culture and cognition* (pp. 15–33). Cambridge, England: Cambridge University Press.

Nisbett, R.E., Peng, K., Choi, I., & Norenzayan, A. (in press). (See References)

Sperber, D., Premack, D., & Premack, A.J. (Eds.). (1995). *Causal cognition: A multidisciplinary debate.* Oxford, England: Oxford University Press.

Note

1. Address correspondence to Richard E. Nisbett, Department of Psychology, University of Michigan, Ann Arbor, MI 48109; e-mail: nisbett@umich.edu.

References

Choi, I., Nisbett, R.E., & Norenzayan, A. (1999). Causal attribution across cultures: Variation and universality. *Psychological Bulletin, 125*, 47–63.

Dweck, C.S., Hong, Y.-Y., & Chiu, C.-Y. (1993). Implicit theories: Individual differences in the likelihood and meaning of dispositional inference. *Personality and Social Psychology Bulletin, 19*, 644–656.

Gardner, H. (1985). *The mind's new science.* New York: Basic Books.

Ji, L., Peng, K., & Nisbett, R.E. (in press). Culture, control, and perception of relationships in the environment. *Journal of Personality and Social Psychology.*

Jones, E.E., & Harris, V.A. (1967). The attribution of attitudes. *Journal of Experimental Social Psychology, 3*, 1–24.

Masuda, T., & Nisbett, R.E. (1999). *Culture and attention to object vs. field.* Unpublished manuscript, University of Michigan, Ann Arbor.

Miller, J.G. (1984). Culture and the development of everyday social explanation. *Journal of Personality and Social Psychology, 46*, 961–978.

Morris, M.W., & Peng, K. (1994). Culture and cause: American and Chinese attributions for social and physical events. *Journal of Personality and Social Psychology, 67*, 949–971.

Nakamura, H. (1988). *The ways of thinking of eastern peoples.* New York: Greenwood Press. (Original work published 1960)

Nisbett, R.E., Peng, K., Choi, I., & Norenzayan, A. (in press). Culture and systems of thought: Holistic vs. analytic cognition. *Psychological Review.*

Norenzayan, A., Choi, I., & Nisbett, R.E. (1999). *Eastern and Western folk psychology and the prediction of behavior.* Unpublished manuscript, University of Michigan, Ann Arbor.

Ross, L. (1977). The intuitive psychologist and his shortcomings. In L. Berkowitz (Ed.), *Advances in experimental social psychology* (Vol. 10, pp. 173–220). New York: Academic Press.

Witkin, H.A., Dyk, R.B., Faterson, H.F., Goodenough, D.R., & Karp, S.A. (1974). *Psychological differentiation.* Potomac, MD: Erlbaum.

Critical Thinking Questions

1. When Westerners and East Asians learn about the same events, they sometimes make different causal attributions – Westerners to the person and East Asians to the situation. However, different attributions may also be made by people within the same culture. For example, if a young man from a poor

neighborhood gets into trouble with drugs, liberals might blame the poverty and conservatives might blame the parents. Do these differences mean that people are, in fact, using different reasoning processes or have different cognitive skills? Or might it mean that they are relying on different information when they reason?

2. East Asians and Westerners also differ in field dependence. Does that necessarily imply that they should make different causal attributions?
3. Would you expect East Asians and Westerners to differ on the illusory control experiments described by Thompson? (If you haven't read Thompson, the general experimental paradigm is described in the overview to this section.)

Solving Problems and Making Decisions

The topics in the previous section and in this section—reasoning, problem solving, and decision making—are usually lumped together under the term "higher-order cognition". Discussions of these abilities often presuppose that we have gathered, through our senses or out of memory, some kind of information, and that we are going to consciously, logically, and rationally use it to try to understand events, solve a problem, make a decision, or achieve a goal.

Research from the 1970's–90's is replete with examples of how people use "heuristics"—shortcuts in reasoning that don't always lead to the correct answer—and how people demonstrate "biases"—systematic errors in reasoning. One conclusion that might be drawn from that body of research is that humans are "irrational" in our thinking. However, in the last decade, the debate over what counts as "rational" has heated up. Does rationality imply that we must be consistent with the dictates of formal mathematical or logical reasoning? Or that we make the best decisions we can make given our limited cognitive capacities? Or that we make decisions that are, if not perfect, obviously good enough for us to have survived given the structure of the environment in which our cognition evolved?

In the previous section we looked at two possible influences on causal reasoning—evolution and culture. In this section we look at other possible influences on reasoning including our unconscious, our emotions, and the opinions of people around us. Each influence calls into question exactly what "rationality" entails.

In "What Have Psychologists (and Others) Discovered about the Process of Scientific Discovery?," Klahr and Simon describe how researchers study the scientific discovery process. Although scientific *reasoning* is often portrayed as a logical process (especially in undergraduate Research Methods textbooks), scientific *discovery* is far messier, and may depend on a combination of knowledge, perseverance, fortuity, and insight. That last issue is the theme taken up by Siegler in "Unconscious Insights". Using an anecdote and experimental evidence, he argues that we may sometimes solve problems unconsciously before we become aware of our solution strategy.

The last two articles address influences on decision making. In "Anticipated Emotions as Guides to Choice," Mellers and McGraw describe how people's knowledge of alternative outcomes to an event can affect their emotional reactions to that event. They also describe how and when people are not good at predicting their own future emotions. One implication of those studies is that if we make choices based on our anticipated emotions, we could be making bad decisions quite often.

One way to make better decisions might be to ask other people for input. In "The Benefit of Additional Opinions", Yaniv describes how and when asking others for their opinions will enhance our own decision making.

Taken together these articles address the question: How do we solve problems and make decisions—and what *do* we, and *should* we, rely on as we do so? Afterwards, we are still left with the question: are humans "rational" (and by what definition) in these—our "highest-order"—endeavors?

What Have Psychologists (And Others) Discovered About the Process of Scientific Discovery?

David Klahr[1] and Herbert A. Simon
Department of Psychology, Carnegie Mellon University, Pittsburgh, Pennsylvania

Abstract

We describe four major approaches to the study of science—historical accounts of scientific discoveries, psychological experiments with nonscientists working on tasks related to scientific discoveries, direct observation of ongoing scientific laboratories, and computational modeling of scientific discovery processes—by viewing them through the lens of the theory of problem solving. We compare and contrast the different approaches, indicate their complementarities, and provide examples from each approach that converge on a set of principles of scientific discovery.

Keywords

scientific discovery; problem solving

Early in the 20th century, Einstein, in reflecting on his own mental processes leading to the theory of relativity, said, "I am not sure whether there can be a way of really understanding the miracle of thinking" (Wertheimer, 1945, p. 227). However, in the past 25 years, several disciplines, including psychology, history, and artificial intelligence,[2] have produced a substantial body of knowledge about the process of scientific discovery that allows us to say a great deal about it.[3] Although the strengths of one approach are often the weaknesses of another, the work has collectively yielded consistent insights into the scientific discovery process.

ASSESSING THE FOUR MAJOR APPROACHES

Historical accounts of the great scientific discoveries—typically based on diaries, scientific publications, autobiographies, lab notebooks, correspondence, interviews, grant proposals, and memos of famous scientists—have high face validity. That is, it is clear that they are based on what they purport to study: real science. However, such studies have some weaknesses. For one thing, their sources are often subjective and unverifiable. Moreover, the temporal resolution of historical analysis is often coarse, but it can become much finer when laboratory notebooks and correspondence are available. Historical investigations often generate novel results about the discovery process, by focusing on a particular scientist and state of scientific knowledge, as well as by highlighting social and motivational factors not addressed by other approaches.

Although historical studies of discovery focus much more on successes than on failures, laboratory studies are designed to manipulate the discovery context in order to examine differences in processes associated with success and failure. Face validity of lab studies varies widely: from studies only distantly related to real scientific tasks to those that model essential aspects of specific scientific discoveries (e.g., Dunbar's, 1993, simulated molecular genetics laboratory; Schunn & Anderson's, 1999, comparison of experts' and novices' ability to design and interpret memory experiments; Qin & Simon's, 1990, study in which college sophomores rediscovered Kepler's third law of planetary motion). Laboratory studies tend to generate fine-grained data over relatively brief periods and typically ignore or minimize social and motivational factors.

The most direct way to study science is to study scientists in their day-to-day work, but this is extraordinarily difficult and time-consuming. A recent example is Dunbar's (1994) analysis of discovery processes in several world-class molecular genetics research labs. Such studies have high face validity and potential for detecting new phenomena. Moreover, they may achieve much finer-grained temporal resolution of ongoing processes than historical research, and they provide rigor, precision, and objectivity that is lacking in retrospective accounts.

A theory of discovery processes can sometimes be incorporated in a computational model that simulates and reenacts discoveries. Modeling draws upon the same kinds of information as do historical accounts, but goes beyond history to hypothesize cognitive mechanisms that can make the same discoveries, following the same path. Modeling generates theories and tests them against data obtained by the other methods. It tests the sufficiency of the proposed mechanisms to produce a given discovery and allows comparison between case studies, interpreting data in a common language to reveal both similarity and differences of processes. Modeling enables us to express a theory rigorously and to simulate phenomena at whatever temporal resolution and for whatever durations are relevant.

SCIENTIFIC DISCOVERY AS PROBLEM SOLVING

Crick argued that discoveries are major when they produce important knowledge, whether or not they employ unusual thought processes: "The path [to the double helix]. . . was fairly commonplace. What was important was *not the way it was discovered*, but the object discovered—the structure of DNA itself" (Crick, 1988, p. 67; italics added). Psychologists have been making the case for the "nothing special" view of scientific thinking for many years. This does not mean that anyone can walk into a scientist's lab and make discoveries. Practitioners must acquire an extensive portfolio of methods and techniques, and must apply their skills aided by an immense base of shared knowledge about the domain and the profession. These components of expertise constitute the *strong methods*. The equally important *weak methods* scientists use underlie all human problem-solving processes.

A problem consists of an initial state, a goal state, and a set of operators for transforming the initial state into the goal state by a series of intermediate steps. Operators have constraints that must be satisfied before they can be applied. The set of states, operators, and constraints is called a *problem space*, and the problem-solving process can be characterized as a search for a path that links initial state to goal state.

Initial state, goal state, operators, and constraints can each be more or less well-defined. For example, one could have a well-defined initial state and an ill-defined goal state and set of operators (e.g., make "something pretty" with these materials and tools), or an ill-defined initial state and a well-defined final state (e.g., find a vaccine against HIV). But well-definedness depends on the familiarity of the problem-space elements, and this, in turn, depends on an interaction between the problem and the problem solver.

Although scientific problems are much less well-defined than the puzzles commonly studied in the psychology laboratory, they can be characterized in these terms. In both cases, well-definedness and familiarity depend not only on the problem, but also on the knowledge that is available to the scientist. For that reason, much of the training of scientists is aimed at increasing the degree of well-definedness of problems in their domain. The size of a problem space grows exponentially with the number of alternatives at each new step in the problem (e.g., the number of possible paths one must consider at each possible move when planning ahead in chess). Effective problem solving must constrain search to only a few such paths. Strong methods, when available, find solutions with little or no search. For example, in chess, there are many standard openings that allow experts to make their initial moves with little search. Similarly, someone who knows algebra can use simple linear equations to choose between two sets of fixed and variable costs when deciding which car to rent instead of painstakingly considering the implications of driving each car a different distance. But the problem solver must first recognize the fit between the given problem (renting a car) and the strong method (high school algebra).

Weak methods, requiring little knowledge of a problem's structure, do not constrain search as much. One particularly important weak method is analogy, which attempts to map a new problem onto one previously encountered, so that the new problem can be solved by a known procedure. However, the mapping may be quite complex, and it may fail to produce a solution.

Analogy enables the problem solver to shift the search from the given problem space to one in which the search may be more efficient, sometimes making available strong methods that greatly abridge search. Prior knowledge can then be used to plan the next steps of problem solving, replace whole segments of step-by-step search, or even suggest an immediate solution. The recognition mechanism uses this store of knowledge to interpret new situations as instances of previously encountered situations. This is a key weapon in the arsenal of experts and a principal factor in distinguishing expert from novice performance.

In the past 25 years, analogy has assumed prominence in theories of problem solving and scientific discovery. Nersessian (1984) documented its role in several major 19th-century scientific discoveries. Recent studies of contemporary scientists working in their labs have revealed the central role played by analogy in scientific discovery (Dunbar, 1994; Thagard, 1998).

Although many strong methods are applied in scientific practice, weak methods are of special interest for scientific discovery because they are applicable in a wide variety of contexts, and strong methods become less available as the scientist approaches the boundaries of knowledge.

COMPLEMENTARITY OF APPROACHES

Viewing scientific discovery as problem solving provides a common language for describing it and facilitates studying the same discovery using more than one approach.

In the late 1950s, Monod and Jacob discovered how control genes regulate the synthesis of lactose (a sugar found in milk) in bacteria (Jacob & Monod, 1961). The literature explaining this discovery (e.g., Judson, 1996) tends to use terms such as "a gleam of perception," but to characterize a discovery as a gleam of perception is to not describe it at all. One must identify specific and well-understood cognitive processes and then determine their role in the discovery. Among the most important steps for Jacob and Monod in discovering the mechanisms of genetic control were representational changes that enabled them to replace their entrenched idea—that genetic control must involve some kind of activation—with the idea that it employed inhibition instead.

Dunbar (1993) created a laboratory task that captured important elements of Monod and Jacob's problem, while simplifying to eliminate many others. His participants—asked to design and run (simulated) experiments to discover the lactose control mechanism—faced a real scientific task with high face validity. Although the task was simplified, the problem, the "givens," the permissible research methods, and the structure of the solution were all preserved. Dunbar's study cast light on the problem spaces that Monod and Jacob searched, and on some of the conditions of search that were necessary or sufficient for success (e.g., knowing that there was such a thing as a control gene, but not exactly how it worked).

In this example, a historically important scientific discovery provided face validity for the laboratory study, and the laboratory provided information about the discovery processes with fine-grained temporal resolution.

CONVERGENT EVIDENCE OF DISCOVERY PRINCIPLES

In this section, we give a few additional examples of convergent evidence obtained by using two or more approaches to study the same discovery.

Surprise

Recently, reigning theories of the scientific method have generally taken hypotheses as unexplained causes that motivate experiments designed to test them. In this view, the hypotheses derive from scientists' "intuitions," which are beyond explanation. Historians of science have taken a less rigid position with respect to hypotheses, and include their origins within the scope of historical inquiry.

For example, the discovery of radium by the Curies started with their attempt to obtain pure radioactive uranium from pitchblende. As they proceeded, they were surprised to find in pitchblende levels of radioactivity higher than in pure uranium. As a surprise calls for an explanation, they conjectured that the pitchblende contained a second substance (which they named radium) more radioactive than uranium. They succeeded in extracting the radium and determined its key properties. In this case, a phenomenon led to a hypothesis, rather than a hypothesis leading to an experimental phenomenon. This occurs frequently in science. A surprise violates prior expectations. In the face of surprise, scientists frequently divert their path to ascertain the scope and import of the surprising phenomenon and its mechanism.

Response to surprise was investigated in a laboratory study (Klahr, Fay, & Dunbar, 1993) in which participants had to discover the function of an unknown key on a simulated rocket ship. They were given an initial hypothesis about how the key worked. Some participants were given a plausible hypothesis, and others were given an implausible hypothesis. In all cases, the suggested hypothesis was wrong, and the rocket ship produced some unexpected, and sometimes surprising, behavior. Adults reacted to an implausible hypothesis by proposing a competing hypothesis and then generating experiments that could distinguish between them. In contrast, children (third graders) tended to dismiss an implausible hypothesis and ignore evidence that supported it. Instead, they attempted to demonstrate that their favored hypothesis was correct. It seems that an important step in acquiring scientific habits of thinking is coming to accept, rather than deny, surprising results, and to explore further the phenomenon that gave rise to them.

Krebs's biochemical research leading to the discovery of the chain of reactions (the reaction path) by which urea (the end product of protein metabolism) is synthesized in the body has been the topic of convergent studies focusing on response to unexpected results. The discovery has been studied historically by Holmes (1991) and through the formulation of two computational models (Grasshoff & May, 1995; Kulkarni & Simon, 1988), both of which have modeled the discovery. After the models proposed an experiment and were given its outcome, they then proposed another experiment, using the previous outcomes to guide their decision about what sort of experiment would be useful. Using no more knowledge of biochemistry than Krebs possessed at the outset, both programs discovered the reaction path by following the same general lines of experimentation as Krebs followed. One of these models (Kulkarni & Simon, 1988) addressed the surprise issue directly (in this case, surprise in finding a special catalytic role for the amino acid ornithine). The simulated scientist formed expectations (as did Krebs) about experimental outcomes. When the expectations were violated, steps were taken to explain the surprise. Thus, historical studies, simulation models, and laboratory experiments all provide evidence that the scientist's reaction to phenomena—either observational or experimental—that are surprising can lead to generating and testing new theories.

Multiple Search Spaces

This reciprocal relation between hypotheses and phenomena arises in laboratory studies, historical studies, and computational models of discovery, enabling us to characterize scientists' thinking processes as problem-solving search in multiple spaces.

Dual Search

The discovery process can be characterized as a search in two spaces: a hypothesis space and an experiment space. When attempting to discover how a particular control button worked on a programmable device, participants in the "rocket ship" study described earlier (Klahr & Dunbar, 1988) had to negotiate this dual search by (a) designing experiments to disclose the button's functions (searching the experiment space) and (b) proposing rules that explained the device's behavior (searching the hypothesis space). Thus, participants were required to coordinate two kinds of problems, and they approached this dual search with different emphases. Some ("experimenters") focused on the space of possible manipulations, whereas others ("theorists") focused on the space of possible explanations of the responses.

Historical studies usually reveal both hypothesis-space search and experiment-space search. For example, most histories of Faraday's discovery of induction of electricity by magnets place much emphasis on the influence of Ampère's theory of magnetism on Faraday's thought. However, a strong case can be made that Faraday's primary search strategy was in the space of experiments, his discovery path being shaped by phenomena observed through experimentation more than by theory.

The number of search spaces depends on the nature of the scientific problem. For example, in describing the discovery of the bacterial origins of stomach ulcers, Thagard (1998) demonstrated search in three major spaces: hypothesis space, experiment space, and a space of instrumentation.

Analogy in Search for Representations

Bohr used the solar system analogy to arrive at his quantum model of the hydrogen atom. He viewed the electrons in the hydrogen atom as planets orbiting the nucleus, although, according to classical understanding of the solar system, this would mean that the charged electrons would dissipate energy until they fell into the nucleus. Instead of abandoning the analogy, Bohr borrowed Planck's theory that energy could be dissipated only in quantum leaps, then showed that these leaps would produce precisely the spectrum of light frequencies that scientists 30 years previously had demonstrated hydrogen produces when its electrons move from a higher-energy stationary state to a lower-energy one.

Search in the Strategy Space

Finally, changes in strategy, even while the representation of a problem is fixed, may enable discovery. Often the change in strategy results from, or leads to, the invention of new scientific instruments or procedures. Breeding experiments go back to Mendel (and experiments for stock breeding go much further back), but the productivity of such experiments depended on mutation rates. Müller, with the "simple" idea that x-rays could induce higher rates of mutation, substantially raised that productivity.

CREATIVITY AND PROBLEM SOLVING IN SCIENCE AND BEYOND

Scientific discovery is a type of problem solving using both weak methods that are applicable in all disciplines and strong methods that are mainly domain-specific. Scientific discovery is based on heuristic search in problem spaces: spaces of instances, of hypotheses, of representations, of strategies, of instruments, and perhaps others. This heuristic search is controlled by general mechanisms such as trial-and-error, hill-climbing, means-ends analysis, analogy, and response to surprise. Recognition processes, evoked by familiar patterns in phenomena, access knowledge and strong methods in memory, linking the weak methods to the domain-specific mechanisms.

All of these constructs and processes are encountered in problem solving wherever it has been studied. A painter is not a scientist; nor is a scientist a lawyer or a cook. But they all use the same weak methods to help solve their respective problems. When their activity is described as search in a problem space, each can understand the rationale of the other's activity, however abstruse and arcane the content of any special expertise may appear.

At the outer boundaries of creativity, problems become less well structured, recognition becomes less able to evoke prelearned solutions or domain-specific search heuristics, and more reliance has to be placed on weak methods. The more creative the problem solving, the more primitive the tools. Perhaps this is why "childlike" characteristics, such as the propensity to wonder, are so often attributed to creative scientists and artists.

Recommended Reading

Klahr, D. (2000). *Exploring science: The cognition and development of discovery processes.* Cambridge, MA: MIT Press.

Klahr, D., & Simon, H.A. (1999). Studies of scientific discovery: Complementary approaches and convergent findings. *Psychological Bulletin, 125*, 524–543.

Zimmerman, C. (2000). The development of scientific reasoning skills. *Developmental Review, 20*, 99–149.

Acknowledgments—Preparation of this article and some of the work described herein were supported in part by a grant from the National Institute of Child Health and Human Development (HD 25211) to the first author. We thank Jennifer Schnakenberg for a careful reading of the penultimate draft.

Notes

1. Address correspondence to David Klahr, Department of Psychology, Carnegie Mellon University, Pittsburgh, PA 15213.

2. In addition, the sociology of science explains scientific discovery in terms of political, anthropological, or social forces. The mechanisms linking such forces to scientific practice are usually motivational, social-psychological, or psychodynamic, rather than cognitive. Although this literature has provided important insights on how social and professional constraints influence scientific practices, we do not have much to say about it in this brief article.

3. This article summarizes an extensive review listed as the second recommended reading. Full references to historical sources alluded to in the present article can be found there, as well as in the first recommended reading. The third recommended reading focuses on developmental aspects of the discovery process.

References

Crick, F. (1988). *What mad pursuit: A personal view of scientific discovery.* New York: Basic Books.

Dunbar, K. (1993). Concept discovery in a scientific domain. *Cognitive Science, 17*, 397–434.

Dunbar, K. (1994). How scientists really reason: Scientific reasoning in real-world laboratories. In R.J. Sternberg & J. Davidson (Eds.), *The nature of insight* (pp. 365–395). Cambridge, MA: MIT Press.

Grasshoff, G., & May, M. (1995). From historical case studies to systematic methods of discovery: Working notes. In *American Association for Artificial Intelligence Spring Symposium on Systematic Methods of Scientific Discovery* (pp. 45–56). Stanford, CA: AAAI.
Holmes, F.L. (1991). *Hans Krebs: The formation of a scientific life, 1900–1933, Volume 1*. New York: Oxford University Press.
Jacob, F., & Monod, J. (1961). Genetic regulatory mechanisms in the synthesis of proteins. *Journal of Molecular Biology, 3*, 318–356.
Judson, H.F. (1996). *The eighth day of creation: Makers of the revolution in biology* (expanded ed.). Plainview, NY: Cold Spring Harbour Laboratory Press.
Klahr, D., & Dunbar, K. (1988). Dual space search during scientific reasoning. *Cognitive Science, 12*, 1–55.
Klahr, D., Fay, A.L., & Dunbar, K. (1993). Heuristics for scientific experimentation: A developmental study. *Cognitive Psychology, 24*, 111–146.
Kulkarni, D., & Simon, H.A. (1988). The process of scientific discovery: The strategy of experimentation. *Cognitive Science, 12*, 139–176.
Nersessian, N.J. (1984). *Faraday to Einstein: Constructing meaning in scientific theories*. Dordrecht, The Netherlands: Martinus Nijhoff.
Qin, Y., & Simon, H.A. (1990). Laboratory replication of scientific discovery processes. *Cognitive Science, 14*, 281–312.
Schunn, C.D., & Anderson, J.R. (1999). The generality/specificity of expertise in scientific reasoning. *Cognitive Science, 23*, 337–370.
Thagard, P. (1998). Ulcers and bacteria: I. Discovery and acceptance. *Studies in the History and Philosophy of Biology and Biomedical Science, 9*, 107–136.
Wertheimer, M. (1945). *Productive thinking*. New York: Harper & Row.

Critical Thinking Questions

1. This article describes four major approaches to the study of scientific discovery. What are the strengths and weaknesses of each of these techniques? Why is it important to have multiple techniques?
2. What distinguishes a true scientific breakthrough from the kinds of problem solving that we do in everyday life (e.g., discovering a shortcut to school, fixing your computer, etc.)—is it the thought processes involved or the product created?
3. How does the "difficulty" of a problem depend on both the structure of the problem and the knowledge of the problem solver?
4. Many popular games, like Clue, Guess Who, Mastermind, and 20 Questions, involve reasoning that is similar to scientific reasoning. For the games you know—what is the hypothesis space and what is the experiment space? Could these games be used to study scientific reasoning in the laboratory? Why might there be people who say they "hate science" but who still enjoy playing these kinds of games?

Unconscious Insights

Robert S. Siegler[1]

Psychology Department, Carnegie Mellon University, Pittsburgh, Pennsylvania

Abstract

From early in the history of psychology, theorists have argued about whether insights are initially unconscious or whether they are conscious from the start. Empirically identifying unconscious insights has proven difficult, however: How can we tell if people have had an insight if they do not tell us they have had one? Fortunately, although obtaining evidence of unconscious insights is difficult, it is not impossible. The present article describes an experiment in which evidence of unconscious insights was obtained. Almost 90% of second graders generated an arithmetic insight at an unconscious level before they were able to report it. Within five trials of the unconscious discovery, 80% of the children made the discovery consciously, as indicated by their verbal reports. Thus, the initial failure to report the insight could not be attributed to the children lacking the verbal facility to describe it. The results indicate that at least in some cases, insights arise first at an unconscious level, and only later become conscious. Rising activation of the new strategy may be the mechanism that leads children to become conscious of using it.

Keywords

insight; discovery; conscious; unconscious; arithmetic

Do unconscious ideas drive our conscious perceptions, thoughts, and behavior? Over the past decade, advances in cognitive neuropsychology have helped spark a rebirth of interest in this enduring question. To cite one example, there have been documented cases of blindsight, in which patients who have suffered brain damage are unaware of seeing objects but can accurately "guess" the objects' locations (Weiskranz, 1997). The influence of unconscious knowledge on behavior is not limited to brain-damaged patients. College students who are asked to learn artificial grammars, four-way statistical interactions, and other complex rule systems often are unaware of having learned the rules, yet they can use the rule systems to classify novel instances (Lewicki, Hill, & Czyzewska, 1992). Recent findings indicate that even insightful discoveries sometimes arise unconsciously before they reach a conscious (i.e., reportable) level. This article describes some of the research leading to this conclusion.

PERSPECTIVES ON INSIGHT AND CONSCIOUSNESS

Long before there was a scientific field of psychology, philosophers, mathematicians, and scientists described the role of consciousness in their own insights. Archimedes' experience of stepping into the bath, seeing the water rise, and exclaiming "Eureka" is probably the prototypical insight: a sudden change from not knowing a problem's solution to knowing it consciously.

In the Archimedes anecdote, an external event stimulated the insight. Other thinkers have emphasized the contribution of unconscious processes and dreams to their discoveries. Perhaps the prototypical account of this type is Kekule's dream of intertwined snakes, which led him to "see" the structure of the benzene ring.

Although these two cases differ in what led up to the insight, the insight itself emerged suddenly in both cases. Other accounts differ, though. Wittgenstein (1969), for example, compared generation of new ideas to a sunrise: Although our experience is that the new day suddenly "dawns," the amount of light actually grows continuously over a fairly protracted period of time.

These examples suggest two major questions regarding the relation between consciousness and insight: Do insights arise at an unconscious (i.e., nonreportable) level before they arise consciously, and do insights arise suddenly or gradually? These are basic questions about human nature, and they have motivated considerable theorizing over the past century (see Sternberg & Davidson, 1995, and the special section of *American Psychologist* edited by Loftus, 1992, for incisive discussions of classical and current perspectives on these issues). However, the questions have proven resistant to empirical resolution. The main reason is the difficulty of obtaining evidence regarding unconscious insights. Simply put, how can we know that people have an insight if they do not tell us that they had it? Recently, however, Elsbeth Stern and I found a way to obtain independent measures of conscious and unconscious insights and thus to examine the relation between them (Siegler & Stern, 1998).

THE INVERSION TASK

On problems of the form A + B - B (e.g., 18 + 24 - 24), the answer always is A. Such inversion problems are useful for studying insight because they can be solved in either insightful or noninsightful ways. The noninsightful way is to use the standard procedure of adding the first two numbers and subtracting the third. The insightful way involves simply saying the first number.

In addition to allowing both insightful and noninsightful solutions, the inversion task has the unusual property of allowing independent measurement of conscious and unconscious versions of the insight. Immediately retrospective verbal reports provided the measure of conscious use of the insight in the research Stern and I conducted. Young school-age children typically report their arithmetic strategies quite accurately, as indicated by converging evidence from reaction time and error patterns (cf. Siegler, 1987). What made the inversion problems special, however, was that an implicit measure of strategy use, one that did not require any verbal report, also could be obtained: the child's solution time. Ordinarily, solution times are insufficient to infer the strategy that was used on an individual trial. However, they are considerably more useful for inferring strategy use on inversion problems. The reason is that solving the problems via computation generates much slower solution times than solving them by using the arithmetic insight. Consistent with this view, solution times on inversion problems in our study were bimodally distributed: 92% of times were either fast (4 s or less) or slow (8 s or more). Converging evidence from overt behavior supported the view that the fast times reflected use of the insight and that the slow times reflected use of computation. Overt computational activity was observed on 80% of trials classified as computation, versus 0% of trials on which children were classified as using the shortcut. (Methods used to classify strategy use are discussed in the next section.)

CONSCIOUS AND UNCONSCIOUS DISCOVERIES

Having both a verbal report and a solution time on each trial made it possible to define three main strategies: *computation*, *shortcut*, and *unconscious shortcut*. Children were classified as using the computation strategy on each trial on which they took more than 4 s to come up with a solution and said they computed the answer; they were classified as using the shortcut on each trial on which the solution time was 4 s or less and they said they used the shortcut; and they were classified as using the unconscious shortcut on each trial on which their solution time was 4 s or less but they claimed to have computed the answer.

We expected that on the large majority of trials, the verbal-report and solution-time measures of strategy use would converge: Children would either solve the problem quickly and say they used the shortcut or take longer to solve it and say they computed the answer. However, we also expected that sometimes the measures would diverge: The child would solve the problem in 4 s or less but claim to have solved the problem through addition and subtraction. Such trials, if they occurred most often at predicted places in the learning sequence, would indicate unconscious use of the shortcut.

THE UNCONSCIOUS-ACTIVATION HYPOTHESIS

Based on previous research showing unconscious influences on other types of thinking, we formulated the *unconscious-activation hypothesis*: Increasing activation of a strategy leads to people first using it unconsciously; then, as the activation increases further, people become conscious of using it. The straightforward implication of this hypothesis was that the unconscious shortcut would emerge before the conscious version of the strategy.

To further test the unconscious-activation hypothesis, we created two experimental conditions. One was the *blocked-problems condition*. Children in it were presented A + B - B problems, that is, problems that could be solved by the inversion principle, on 100% of the trials. The other experimental condition was the *mixed-problems condition*. In it, children were presented A + B - B problems on 50% of trials, and on the other 50% were presented A + B - C problems (i.e., problems, such as 18 + 24 - 15, in which the three numbers differed and therefore that could not be solved via the shortcut strategy). The unconscious-activation hypothesis predicted that presenting children inversion problems on 100% of the trials would lead to a more rapid increase in activation of the shortcut, which in turn would lead to (a) more rapid discovery of the unconscious-shortcut and shortcut strategies (discovery after fewer inversion problems), (b) a shorter gap between discovery of the unconscious shortcut and discovery of the shortcut, (c) more consistent use of the shortcut on inversion problems once it was discovered, and (d) greater generalization of the strategy to novel problems of similar appearance, such as A - B + B and A + B + B.

AN EXPERIMENT ON CONSCIOUS AND UNCONSCIOUS INSIGHTS

To test these predictions, we presented 31 German second graders with either the blocked problems or the mixed problems. The experiment was conducted over an 8-week period, one session per week.

Each of the predictions of the unconscious-activation hypothesis was borne out. Almost 90% of the children discovered the unconscious version of the shortcut before the conscious version. Relative to children in the mixed-problems condition, children in the blocked-problems condition discovered both the unconscious-shortcut and the shortcut strategies after seeing fewer inversion problems, exhibited a shorter gap between their discovery of the two strategies, used the strategies more often once they discovered them, and generalized the strategies more widely to novel types of problems.

Examination of strategy use just before and after discovery of the unconscious-shortcut and shortcut strategies provided particularly direct support for the unconscious-activation hypothesis. Figure 1 illustrates the circumstances surrounding the first use of the unconscious shortcut among children in the blocked-problems condition. Trial 0 for a given child is the trial on which the

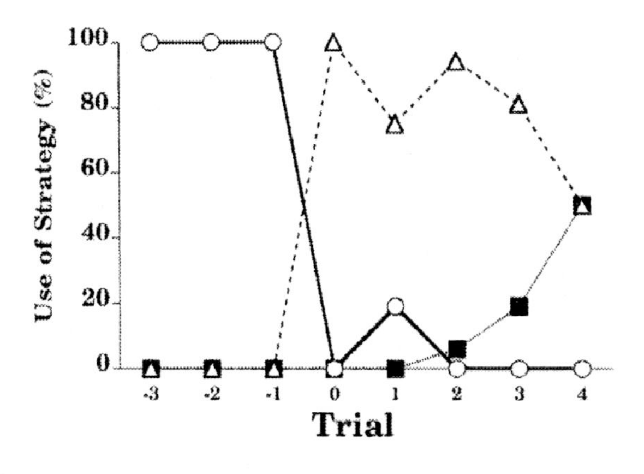

Fig. 1. Percentage use of computation, unconscious-shortcut, and shortcut strategies in the blocked-problems condition on trials immediately preceding and following children's first use of the unconscious shortcut. Each child's first use of the unconscious shortcut is designated Trial 0; the trial just before it is designated Trial -1, the trial just after it is designated Trial 1, and so on.

child first used the unconscious shortcut; thus, by definition, 100% of children used the unconscious shortcut on Trial 0. Trial -1 for a given child is whichever trial immediately preceded that child's Trial 0; Trial 1 is whichever trial immediately followed the child's Trial 0; and so on.

Data from the blocked-problems condition were particularly striking. Figure 1 reveals that just before their first use of the unconscious shortcut, all of these children used the computation strategy. After their initial use of the unconscious shortcut, most of them continued to use the unconscious shortcut over the next three trials. By the fourth trial after the initial use of the unconscious shortcut, half of the children reported using the shortcut. By the fifth trial, 80% of the children reported using it.

Figure 2 shows a parallel analysis centered on first use of the shortcut by the same children. On the three trials immediately preceding its first use, roughly 80% of these children used the unconscious shortcut (as opposed to less than 10% use of this strategy for the study as a whole). Once the children began to report using the shortcut, they continued to use it quite consistently within that session. However, when they returned a week later for the next session, fewer than 35% used the shortcut on any trial before Trial 5. Thereafter, more children rediscovered the shortcut, and by the end of the session, more than 90% of them were again using it.

Changes in solution times from the trials just before the first use of the unconscious shortcut to the trial of discovery suggested that the unconscious shortcut represented a sudden, qualitative shift in thinking. On the three trials immediately before its first use, solution times averaged 12 s; on its first use, the mean solution time was 2.7 s. Solution times on subsequent unconscious-shortcut trials (and on

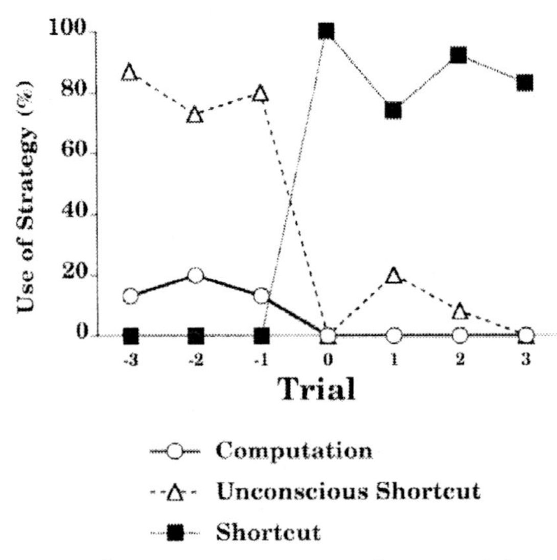

Fig. 2. Percentage use of computation, unconscious-shortcut, and shortcut strategies in the blocked-problems condition on trials immediately preceding and following children's first use of the shortcut. Each child's first use of the shortcut is designated Trial 0; the trial just before it is designated Trial -1, the trial just after it is designated Trial 1, and so on.

shortcut trials as well) continued to average between 2 s and 3 s in all sessions. Thus, although children who used the unconscious shortcut did not report doing anything different, they had already had the insight at a behavioral level.

The lack of reporting of the insight on unconscious-shortcut trials could not be attributed to the children being generally inarticulate, to the insight being difficult to put into words, or to children's perceptions of social desirability preventing them from reporting an approach that they knew they were using. If those were the reasons for children initially not reporting the shortcut, why would the same children have almost invariably reported using it a few trials later in the same session? Further supporting the view that use of the shortcut was at first unconscious, when children rediscovered the shortcut in the session following the one in which they initially used it, most again used the unconscious version just before beginning to report its use.

CONCLUSIONS

These results shed light on both of the questions raised at the outset regarding insights and consciousness. With regard to the first question, the findings demonstrate that insights are not always conscious from the start. At least sometimes, they arise first in unconscious form.

The results also provide an answer to the second question: Insights are abrupt in some senses, but gradual in others. On the one hand, the dramatic reduction in solution times that accompanied the first use of the unconscious shortcut indicates a sense in which insight was abrupt. The fact that solution times on shortcut trials did not decline further indicates that in terms of efficiency of execution, the shortcut emerged full-blown. On the other hand, the insight was gradual in two other senses. First, children initially discovered the shortcut in a nonreportable form and only later became able to report using it. Second, use of the shortcut increased slowly, never extending to more than 60% of trials in a given session.

The results also raise several intriguing questions. Do adults also begin to use new strategies unconsciously before they become conscious of using them, or is unconscious use of strategies unique to children? Are unconscious insights limited to single-step strategies, as in the present case, or do people also discover multistep strategies at an unconscious level before they discover them consciously? Perhaps most important, through what cognitive processes are unconscious insights generated?

Underlying these and other relatively specific questions is the main point of our study, a point consistent with a wide range of previous research: Having a thought, or even an insight, is not the same as being aware of having that thought or insight. Learning how consciousness is related to insight remains one of the basic challenges in understanding human psychology, just as it was in the days of Archimedes.

Recommended Reading

Davidson, J. (1995). The suddenness of insight. In R.J. Sternberg & J.E. Davidson (Eds.), *The nature of insight* (pp. 125–155). Cambridge, MA: MIT Press.

Goldin-Meadow, S., Alibali, M.W., & Church, R.B. (1993). Transitions in concept acquisition: Using the hand to read the mind. *Psychological Review, 100*, 279–297.
Schooler, J.W., Fallshore, M., & Fiore, S.M. (1995). Epilogue: Putting insight into perspective. In R.J. Sternberg & J.E. Davidson (Eds.), *The nature of insight* (pp. 367–402). Cambridge, MA: MIT Press.
Siegler, R.S., & Stern, E. (1998). (See References)

Acknowledgments—I would like to thank the National Institutes of Health (Grant 19011) and the Spencer Foundation for funding that helped make possible much of the research reported in this article.

Note

1. Address correspondence to Robert S. Siegler, Psychology Department, Carnegie Mellon University, Pittsburgh, PA 15213.

References

Lewicki, P., Hill, T., & Czyzewska, M. (1992). Nonconscious acquisition of information. *American Psychologist, 47*, 796–801.
Loftus, E.F. (Ed.). (1992). Science watch [Special section]. *American Psychologist, 47*, 761–809.
Siegler, R.S. (1987). The perils of averaging data over strategies: An example from children's addition. *Journal of Experimental Psychology: General, 116*, 250–264.
Siegler, R.S., & Stern, E. (1998). A microgenetic analysis of conscious and unconscious strategy discoveries. *Journal of Experimental Psychology: General, 127*, 377–397.
Sternberg, R.J., & Davidson, J.E. (Eds.). (1995). *The nature of insight*. Cambridge, MA: MIT Press.
Weiskranz, L. (1997). *Consciousness lost and found*. Oxford, England: Oxford University Press.
Wittgenstein, L. (1969). *On certainty*. New York: Harper & Row.

Critical Thinking Questions

1. This article mentions the famous parable about Kekule—that he discovered the structure of benzene after dreaming about a snake. If someone tells you how he solved a problem or why she had a particular insight, should you necessarily believe them (assuming that they are not lying)? In particular, should you believe an account written long after the event occurred? (Recall what you know about the reconstructive nature of memory.)

2. In the Inversion task, the conscious strategy was measured by asking the children which strategy they used and the unconscious strategy was measured by solution time. Given the two conscious categories and two unconscious categories, there should be four combinations. What is each called? Which one doesn't exist? In what order did the children go through them?

3. Do you believe that you have ever used an unconscious strategy before you became aware of it? Think of solving math, statistics, chemistry, or physics problems. Also think of knowing what to do in a card game or where to move in a sport without being taught when and why to take those specific actions.

Anticipated Emotions as Guides to Choice

Barbara A. Mellers[1] and A. Peter McGraw
Department of Psychology, The Ohio State University, Columbus, Ohio

Abstract

When making decisions, people often anticipate the emotions they might experience as a result of the outcomes of their choices. In the process, they simulate what life would be like with one outcome or another. We examine the anticipated and actual pleasure of outcomes and their relation to choices people make in laboratory studies and real-world studies. We offer a theory of anticipated pleasure that explains why the same outcome can lead to a wide range of emotional experiences. Finally, we show how anticipated pleasure relates to risky choice within the framework of subjective expected pleasure theory.

Keywords

anticipated emotions; choice; pleasure

When making decisions, people often anticipate how they will feel about future outcomes and use those feelings as guides to choice. To understand this process, we have investigated the anticipated and actual pleasure of outcomes that follow decisions in laboratory and real-world studies. In this article, we present a theory of anticipated pleasure called decision affect theory and show how it relates to decision making. We claim that when making decisions, people anticipate the pleasure or pain of future outcomes, weigh those feelings by the chances they will occur, and select the option with greater average pleasure.[2]

Imagine a decision maker who is considering two locations for a summer vacation. The first is perfect in all regards—as long as the weather is nice. Unfortunately, the weather is hard to predict. The second location is quite acceptable, and the weather is almost always good. To make a choice, the decision maker anticipates the pleasure of the first vacation assuming good weather and the displeasure of the vacation assuming bad weather. These feelings are weighted by the perceived likelihood of good or bad weather, respectively, and the resulting feelings are combined to obtain an average feeling of anticipated pleasure. The second location is evaluated in the same manner, and the location with greater average pleasure is selected.

We begin by summarizing several studies and then answer three related questions. First, what variables influence anticipated pleasure? Second, how is anticipated pleasure related to choice? And third, how accurately do people anticipate the pleasure of future outcomes?

EXPERIMENTS

In our laboratory studies, we presented participants with pairs of monetary gambles on a computer screen (Mellers, Schwartz, Ho, & Ritov, 1997; Mellers, Schwartz, & Ritov, 1999). Each gamble was displayed as a pie chart with two regions, representing wins or losses. On each trial, respondents chose the gamble they preferred to play. In some conditions, a spinner appeared in the center of the chosen gamble and began to rotate. Eventually it stopped, and participants learned how much they won or lost. In other conditions, spinners appeared in the center of both gambles. The spinners rotated independently and eventually stopped, at which point participants learned their outcome and that of the other gamble. Outcomes ranged from a $32 win to a $32 loss. In some studies, the outcomes were hypothetical, and in others, the outcomes were real. If the outcome was hypothetical, participants anticipated the pleasure they would have felt had the outcome been real. If the outcome was real, participants rated their actual pleasure. Both types of judgments were made on a category rating scale from "very happy" to "very unhappy."

Within this paradigm, participants are likely to find two counterfactual comparisons particularly salient (Bell, 1982, 1985; Loomes & Sugden, 1982, 1986). Comparisons of the imagined outcome with other outcomes of the chosen gamble are called disappointment or elation. Comparisons of the imagined outcome with an outcome of the unchosen gamble are called regret or rejoicing.

In the real-world studies, we used participants who had already made a choice, but did not yet know the outcome of their choice. We asked them to anticipate their feelings about all possible outcomes of the choice. Later, when they learned what the actual outcome was, they rated their feelings regarding what occurred. The studies involved grades, diets, and pregnancy tests. In the grading study, undergraduates predicted their final grade in introductory psychology and anticipated their emotional reactions to all possible grades. The following quarter, they told us their actual grades and feelings about those grades. In the dieting study, clients participating in a commercial weight-loss program told us their weekly weight-loss goals and anticipated their feelings about various outcomes. They returned the following week, learned about their weight changes, and reported their feelings. Finally, in the pregnancy study, women waiting for a pregnancy test at Planned Parenthood anticipated their emotions about possible test results. Ten minutes later, they learned the results and judged their reactions. We now present selected results from these studies.

WHAT VARIABLES INFLUENCE ANTICIPATED PLEASURE?

Our most important findings about pleasure are shown in Figure 1, which presents results from the gambling studies. Our findings can be summarized in terms of outcome effects, comparison effects, and surprise effects. Outcome effects are illustrated in Figure 1a. As the amount of the imagined outcome increases, so does the anticipated pleasure.

Figures 1b and 1c show comparison effects. Figure 1b plots the anticipated pleasure of an obtained win of $8 or loss of $8, separately for trials on which the unobtained outcome was a loss of $32 or gain of $32. When the unobtained outcome was more desirable, the anticipated pleasure about the obtained outcome declined. This is because people anticipate disappointment when they imagine getting the worse outcome of two outcomes. Figure 1c plots the anticipated pleasure of an obtained win of $8 or a loss of $8, separately for trials on

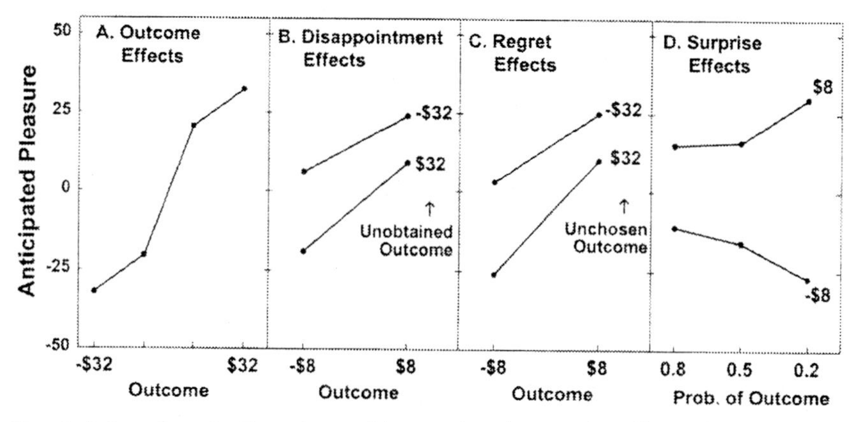

Fig. 1. Selected results from the gambling studies showing the effects of outcomes (a), comparison (b and c), and surprise (d) on anticipated pleasure. Comparison effects are illustrated by anticipated pleasure when the unobtained (b) or unchosen (c) outcome was a loss of \$32 versus a gain of \$32. Surprise effects (d) are shown for both a gain of \$8 and a loss of \$8. Prob. = probability.

which the outcome of the other gamble was a loss or gain of \$32. A similar pattern appears: When the outcome of the unchosen gamble was more appealing, anticipated pleasure decreased. This is because people anticipate regret when they imagine having made the wrong choice.

Comparison effects of both disappointment and regret on anticipated pleasure are asymmetric. The displeasure of getting the worse of two outcomes is typically greater in magnitude than the pleasure of receiving the better outcome. Comparison effects are powerful enough to make an imagined loss that is the better of two losses more pleasurable than an imagined gain that is the worse of two gains, as we found in other studies (Mellers et al., 1999).

The results shown in Figure 1d illustrate the effects of surprise. Participants anticipated more pleasure with a win of \$8 the less likely it was, and they anticipated less pleasure with a loss of \$8 the less likely it was. In other words, both positive and negative feelings are stronger when outcomes are surprising. Surprising outcomes have greater intensity than expected outcomes. Surprise amplifies the emotional experience.

Figure 2 presents selected results from our real-world studies. Figure 2a shows the effects of outcome in the dieting study. As imagined weight loss increased, dieters anticipated greater pleasure. Figure 2b presents the comparison effects in the grading study. Students with lower expectations anticipated greater pleasure from all possible grades. Finally, Figure 2c shows the effects of surprise for women in the pregnancy study who preferred not to be pregnant. Surprising outcomes were associated with more intense anticipated feelings than expected outcomes.

The effects of outcomes, comparisons, and surprise shown in Figures 1 and 2 are predicted by an account of anticipated pleasure called decision affect theory. In this theory, anticipated pleasure depends on the utility (or psycho-

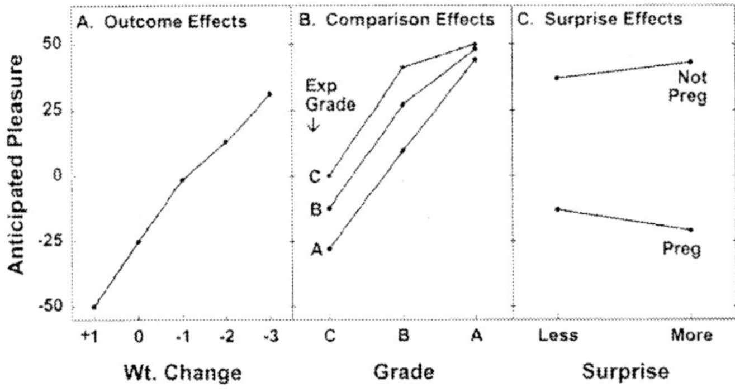

Fig. 2. Selected results from the dieting, grading, and pregnancy studies showing the effects of outcome (a), comparison (b), and surprise (c), respectively, on anticipated pleasure. Comparison effects are illustrated by anticipated pleasure when the expected grade was an A versus a B versus a C. Surprise effects are shown for both women who found out they were pregnant and those who found out they were not pregnant ("Preg"). Wt. = weight.

logical satisfaction) of the outcome and salient comparisons. Comparisons are weighted by how surprising the outcome is. We have provided formal treatments of this theory elsewhere (Mellers et al., 1997, 1999; Mellers & McGraw, 2001).

HOW IS ANTICIPATED PLEASURE RELATED TO CHOICE?

In several studies, we have found that anticipated pleasure is closely connected to choice (Mellers et al., 1999). We assume that decision affect theory predicts the pleasure people anticipate for future outcomes of a given option. Then they weigh those anticipated feelings by the perceived chances of their occurrence, and combine them to form an average anticipated pleasure for each option. The option with greater average pleasure is selected. More detailed descriptions of this theory, called subjective expected pleasure theory, are presented elsewhere (Mellers et al., 1999).

Individuals whose choices are consistent with subjective expected pleasure theory can differ in several respects. For example, if the vacationer we described in the introduction anticipates tremendous pleasure with the first location or is optimistic about good weather, he is more likely to select that location than the alternative location. Greater anticipated pleasure or greater optimism tend to produce greater risk seeking, whereas less anticipated pleasure or more pessimism lead to greater risk aversion.

Subjective expected pleasure theory is similar in some respects to subjective expected utility theory (Savage, 1954). In subjective expected utility theory, decision makers are assumed to consider the utility associated with each outcome, weigh that utility by the perceived chances it will occur, and sum the values for all the outcomes. Utilities are often described in terms of psychological satisfaction, so it seems logical to assume they would not differ from anticipated pleasure. However, utilities do differ from anticipated pleasure. In most theories of choice, utilities depend only on the status quo, but no other reference points. Anticipated pleasure depends on multiple reference points. Furthermore, in most theories of choice, utilities are assumed to be independent of beliefs. In contrast, the anticipated pleasure of outcomes varies systematically with beliefs about their occurrence; anticipated feelings associated with surprising outcomes are amplified relative to anticipated feelings associated with expected outcomes. Because the utility of an outcome differs from the anticipated pleasure of that outcome, the predictions of subjective expected utility theory and subjective expected pleasure theory can differ.

We tested subjective expected pleasure theory by examining whether it could predict participants' actual choices in our gambling studies. To do this, we fit decision affect theory to the anticipated pleasure of outcomes. That is, we estimated parameter values that produced the smallest squared deviations between participants' judgments of anticipated pleasure and the theory's predictions. Then, using these predictions of decision affect theory, we calculated the average anticipated pleasure of each gamble. Finally, we predicted choices based on the assumption that participants select the option with the greater average anticipated pleasure. Predicted choices were correlated with actual choices in five different experiments (Mellers et al., 1999). The correlations ranged from .66 to .86, with an average of .74. These values were remarkably high given the fact that subjective expected pleasure theory was never fit directly to choices. That is, choice predictions were obtained by fitting decision affect theory to judgments of anticipated pleasure.

We further tested decision affect theory by investigating whether anticipated pleasure (which contains utilities, comparisons, and surprise effects) added to the predictability of risky choice over and beyond utilities. To answer this question, we computed the correlations between predicted choices and actual choices after removing the predictions of subjective expected utility theory. These correlations were positive and ranged from .64 to .03, with an average correlation of .33. These analyses show that anticipated pleasure, which is sensitive to comparisons and surprise effects, improves the predictability of choice over and beyond utilities.

HOW WELL CAN DECISION MAKERS ANTICIPATE PLEASURE?

If people make choices by comparing the average anticipated pleasure of options, the accuracy of their predictions becomes a critical concern. Inaccurate predictions could easily lead to peculiar choices. People who overestimate the pleasure of favorable outcomes, for example, would tend to be overly risk seeking. People who overestimate the displeasure of unfavorable outcomes would tend to be overly risk averse.

We examined the accuracy of affective predictions in both laboratory and real-world studies by comparing anticipated pleasure with actual pleasure (Mellers, 2000). In the laboratory studies, predictions were quite accurate. In the pregnancy and dieting studies, however, participants made systematic prediction errors, and those errors were in the same direction. Specifically, participants overestimated the displeasure of unfavorable outcomes. Women who received bad news from their pregnancy tests actually felt better than they expected. These results are surprising because judgments were made only 10 min apart. Likewise, dieters who gained weight or failed to lose it also felt better than they expected. These results are also surprising given the fact that most dieters are quite familiar with attempts to lose weight, and therefore should have experience with their actual reactions to unsuccessful attempts.

Other errors in affective forecasts have also been found (cf. Loewenstein & Schkade, 1999). Errors can occur from the emotions experienced during the choice process. These experienced emotions influence perceptions, memories, and even decision strategies. Other errors occur because people focus on whatever is salient at the moment, what Schkade and Kahneman (1998) call the focusing illusion. In a fascinating demonstration, Schkade and Kahneman asked students at universities in the Midwest and in California to judge their own happiness and the happiness of students at the other location. The comparison highlighted the advantages of California—better climate, more cultural opportunities, and greater natural beauty. Both students in the Midwest and those in California predicted that Californians were happier, but in fact, students at the two locations were equally happy.

The focusing illusion can also lead people to base affective predictions on transitions rather than final states (Kahneman, 2000). Gilbert, Pinel, Wilson, Blumberg, and Wheatley (1998) asked untenured college professors to anticipate how they would feel about receiving or not receiving tenure. Not surprisingly, the professors expected to be happy if given tenure and extremely unhappy otherwise. Actually, however, the professors who were denied tenure were much happier than they expected to be. Errors in affective forecasting that Gilbert et al. found were in the same direction as those we found in the pregnancy and dieting studies. People anticipated feeling worse about negative outcomes than they actually felt.

FUTURE DIRECTIONS

Research in decision making has demonstrated that anticipated pleasure improves the predictability of choice over and beyond utilities. The effects of comparisons and surprise add valuable information to descriptive theories of choice. Disappointment and regret are by no means the only comparisons that influence anticipated pleasure, however. Many other reference points may be salient. When people make a series of gambling choices, for example, the pleasure of a win or loss is affected by previous wins and losses. In competitive situations, people anticipate the pleasure of their success by comparing their performance with that of others, not to mention their own personal expectations.

Many questions remain. Emotions are far more complex than simple unidimensional ratings of pleasure or pain. People can experience pain from sadness, anger, fear, and disappointment. No one would argue that these emotions should be treated as equivalent. Furthermore, some decision outcomes simultaneously give rise to pleasure and pain. In those cases, people feel ambivalence. Finally, what about the duration of emotional experiences? When is regret a fleeting incident, and when does it last a lifetime? Answers to these questions will deepen social scientists' understanding of emotions, and lead to better tools for guiding choice.

Recommended Reading

Gilbert, D.T., & Wilson, T.D. (2000). Miswanting: Some problems in the forecasting of future affective states. In J. Forgas (Ed.), *Thinking and feeling: The role of affect in social cognition* (pp. 178–197). Cambridge, England: Cambridge University Press.

Kahneman, D., & Varey, C. (1991). Notes on the psychology of utility. In J. Elster & J. Roemer (Eds.), *Interpersonal comparisons of well being* (pp. 127–163). New York: Cambridge University Press.

Landman, J. (1993). *Regret: The persistence of the possible.* Oxford, England: Oxford University Press.

Acknowledgments—Support was provided by the National Science Foundation (SBR-94-09819 and SBR-96-15993). We thank Philip Tetlock for comments on an earlier draft.

Notes

1. Address correspondence to Barbara A. Mellers, Department of Psychology, The Ohio State University, Columbus, OH 43210; e-mail: mellers.1@osu.edu; or send e-mail to A. Peter McGraw at mcgraw.27@osu.edu.

2. Pleasure can be derived from acts of virtue, the senses, or relief from pain. Similarly, displeasure can arise from an aggressive impulse, a sense of injustice, or frustration from falling short of a goal. Thus, choices based on pleasure need not imply hedonism.

References

Bell, D.E. (1982). Regret in decision making under uncertainty. *Operations Research, 30*, 961–981.

Bell, D.E. (1985). Disappointment in decision making under uncertainty. *Operations Research, 33*, 1–27.

Gilbert, D.T., Pinel, E.C., Wilson, T.C., Blumberg, S.J., & Wheatley, T.P. (1998). Immune neglect: A source of durability bias in affective forecasting. *Journal of Personality and Social Psychology, 75*, 617–638.

Kahneman, D. (2000). Evaluation by moments: Past and future. In D. Kahneman & A. Tversky (Eds.), *Choices, values, and frames* (pp. 693–708). New York: Cambridge University Press.

Loewenstein, G., & Schkade, D. (1999). Wouldn't be nice? Predicting future feelings. In D. Kahneman, E. Diener, & N. Schwarz (Eds.), *Well-being: The foundations of hedonic psychology* (pp. 85–108). New York: Russell Sage Foundation.

Loomes, G., & Sugden, R. (1982). Regret theory: An alternative of rational choice under uncertainty. *Economic Journal*, 92, 805–824.

Loomes, G., & Sugden, R. (1986). Disappointment and dynamic consistency in choice under uncertainty. *Review of Economic Studies*, 53, 271–282.

Mellers, B.A. (2000). Choice and the relative pleasure of consequences. *Psychological Bulletin.*

Mellers, B.A., & McGraw, A.P. (2001). *Predicting choices from anticipated emotions.* Unpublished manuscript, Ohio State University, Columbus.

Mellers, B.A., Schwartz, A., Ho, K., & Ritov, I. (1997). Decision affect theory: Emotional reactions to the outcomes of risky options. *Psychological Science*, 8, 423–429.

Mellers, B.A., Schwartz, A., & Ritov, I. (1999). Emotion-based choice. *Journal of Experimental Psychology: General, 128*, 332–345.

Savage, L.J. (1954). *The foundations of statistics.* New York: Wiley.

Schkade, D.A., & Kahneman, D. (1998). Does living in California make people happy? *Psychological Science*, 9, 340–346.

Critical Thinking Questions

1. You are at a carnival. You pay one ticket (worth $1) to be able to reach into a dark urn and pull out a chip that indicates whether you have won a prize. You win $3. How happy are you? You say it depends? What does it depend on and why? Is that rational?

   If you have read Spellman & Mandel (from Part 3): How is the answer "it depends" related to issues of counterfactual reasoning and regret?

   (Note: the carnival example reminds us of something the authors tell our students in our statistics classes. Suppose you find out that your friend got a 95 on an exam. You shouldn't offer congratulations until you find out what it was out of—200?—what the mean was—97?—and what the standard deviation was—1? Oops.)

2. Are people good at predicting how they will feel in the future when certain events occur? What kinds of errors do we tend to make? Are we more likely to err in the laboratory or in real life? If you were an insurance salesperson, how could you take advantage of the mispredictions?

3. Emotion is currently a very hot topic in the decision-making literature. One view of emotions is that they lead to mistakes in decision making; another view is that they provide information that can be useful in decision making. Do you think paying attention to emotions when making decisions is "rational"? Why or why not?

The Benefit of Additional Opinions

Ilan Yaniv
Hebrew University of Jerusalem, Jerusalem, Israel

Abstract

In daily decision making, people often solicit one another's opinions in the hope of improving their own judgment. According to both theory and empirical results, integrating even a few opinions is beneficial, with the accuracy gains diminishing as the bias of the judges or the correlation between their opinions increases. Decision makers using intuitive policies for integrating others' opinions rely on a variety of accuracy cues in weighting the opinions they receive. They tend to discount dissenters and to give greater weight to their own opinion than to other people's opinions.

Keywords

judgment and decision making; aggregating opinions; combining information

It is common practice to solicit other people's opinions prior to making a decision. An editor solicits two or three qualified reviewers for their opinions on a manuscript; a patient seeks a second opinion regarding a medical condition; a manager considers several judgmental forecasts of the market before embarking on a new venture. All these situations involve the decision maker in the task of combining other people's opinions, mostly so as to improve the final decision.

People also seek advice when they feel strongly accountable for their decisions. An accountant performing a complex audit might solicit advice to help justify his or her decisions and share the responsibility for the outcome with others. One could justifiably argue, however, that even such reasons for seeking others' opinions are rooted in the belief that this process could improve decision making.

Two main questions arise in the research on combining opinions. One involves the statistical aspects of the combination task: Under what conditions does combining opinions improve decision quality? The other concerns the psychological process of combining judgments: How do judges utilize other people's opinions? These questions, which have been investigated by students of judgment and decision making, statistics, economics, and management, are intertwined, because the quality of the product is related to the way it is produced. In this review, I discuss what researchers have learned about the process and outcomes of combining opinions.

Our focus here is on situations in which a decision maker seeks quantitative estimates, judgments, and forecasts from people possessing the relevant knowledge. The opinions are then combined by the individual decision maker, not by a group (decision making in groups deserves a separate discussion). It is

---

Address correspondence to Ilan Yaniv, Department of Psychology, Hebrew University, Jerusalem 91905, Israel; e-mail: ilan.yaniv@huji.ac.il.

useful to distinguish between two ways in which expert judgments can be combined: (a) intuitively (subjectively) and (b) mechanically (formally), that is, by using a consistent formula, such as simple or weighted averaging.[1]

ACCURACY GAINS FROM AGGREGATION

Research has demonstrated repeatedly that both mechanical and intuitive methods of combining opinions improve accuracy. For example, in a study of inflation forecasts, the aggregate judgment created by averaging the forecasts of expert economists was more accurate than most of these individual forecasts, though not as good as the best ones (Zarnowitz, 1984). The best forecasts, however, could not be identified before the true value became known. Hence, taking the average was superior to selecting the judgment of any of the individuals. Moreover, a small number of opinions (e.g., three to six) is typically sufficient to realize most of the accuracy gains obtainable by aggregation. These fundamental results have been demonstrated in diverse domains, ranging from perception (line lengths) and general-knowledge tasks (historical dates) to business and economics (sales or inflation forecasts), and are an important reason for the broad interest in research on combining estimates (Johnson, Budescu, & Wallsten, 2001; Sorkin, Hayes, & West, 2001; Yaniv & Kleinberger, 2000).

How Does Combining Opinions Improve Judgment?

The improvement in accuracy is grounded in statistical principles, as well as psychological facts. For quantitative estimates, a common measure of accuracy is the average distance of the prediction from the event predicted. In the special case of judgments made on an arbitrary rating scale (e.g., an interviewer's rating of a job candidate's capability on a 9-point scale), a common measure is the correlation between the judgments and some objective outcome (e.g., the candidate's actual success).

In the case of quantitative estimates, it can be outlined in simple terms why improvement is to be expected from combining estimates. A subjective estimate about an objective event can be viewed as the sum of three components: the "truth," random error (random fluctuations in a judge's performance), and constant bias (a consistent tendency to over- or underestimate the event). Statistical principles guarantee that judgments formed by averaging several sources have lower random error than the individual sources on which the averages are based. Therefore, if the bias is small or zero, the average judgment is expected to converge about the truth (Einhorn, Hogarth, & Klempner, 1977).

The case of categorical, binary judgments (e.g., a physician inspects a picture of a tumor and estimates whether it is benign or malignant) requires a special mention. Suppose a decision maker polls the judgments of N independent expert judges whose individual accuracy levels (chances of choosing the correct

---

[1]More complex methods based on Bayes's theorem are less common in psychological research on combining opinions; hence, they are not treated here.

answer) are greater than 50% and then decides according to the majority. For example, three experts might judge whether or not a witness is lying, and the final decision would be the opinion supported by two or more experts. According to a well known 18th-century theorem (known as Condorcet's jury theorem), the accuracy of the majority increases rapidly toward 100% as N increases (e.g., Sorkin et al., 2001). Thus, the majority outperforms the individual judges. For instance, the majority choice of five independent experts who are each correct 65% of the time is expected to be correct approximately 76% of the time.

Conditions Under Which Accuracy Gains Are Observed

A central condition for obtaining optimal accuracy gains through aggregation is that the experts are independent (e.g., little gain is expected if judge B is essentially a replica of judge A). But gains of appreciable size can be observed even when there are low or moderate positive correlations between the judgments of the experts (Johnson et al., 2001). The gains from aggregating quantitative judgments are also determined by the bias and the random error of the estimates (the lower the better). If judgments are made on rating scales, then the accuracy gains are related directly to the validity of each judge (i.e., how the judge's ratings correlate with the objective value of what is rated) and indirectly to the correlations between different judges' ratings (Einhorn et al., 1977; Hogarth, 1978; Johnson et al., 2001).

Number of Opinions Needed

As already noted, as few as three to six judgments might suffice to achieve most of what can be gained from averaging a larger number of opinions. This puzzling result that adding opinions does not contribute much to accuracy is related to my previous comments. Some level of dependence among experts is present in almost any realistic situation (their opinions tend to have some degree of correlation for a variety of reasons—they may rely on similar information sources or have similar backgrounds, or simply consult one another; cf. Soll, 1999). Therefore, the benefits accrued from polling more experts diminish rapidly, with each additional one amounting to "more of the same." Similarly, bias or low judge validity limits the potential accuracy gains and further diminishes the value of added opinions.

PSYCHOLOGICAL EFFECTS ON THE AGGREGATION OF OPINIONS

Consider generic scenarios involving intuitive methods of combining opinions: A moviegoer receives conflicting reviews about a movie, or an undergraduate student hears mixed evaluations from fellow students about an elective course. Although formal approaches deal with the conflict by assigning explicit weights to the various opinions, people often attempt to resolve the conflict by trying to form well-justified, coherent judgments, assessing the merit of each source and the arguments for or against each opinion and trying to explain away the differences. Specifically, several factors affect the weighting of opinions in intuitive decision making, including (a) cues for accuracy, (b) responses to dissension, and (c) self-versus-other effects.

Cues for Accuracy

A decision maker's trust in a given opinion depends on his or her assessment of the accuracy of the source. How are expectations about this accuracy formed? How does trust develop? Studies suggest that a variety of cues serve as proxy measures of the actual accuracy of sources. These cues include expertise, confidence, and past performance.

First, people are sensitive to the expertise (or credibility) ascribed to various sources and assign weights to sources as a function of such attributions (Birnbaum & Stegner, 1979). Second, a frequent and immediate cue for accuracy is the judge's stated confidence about his or her opinion (Sniezek & Van Swol, 2001). Subjective statements such as "Trust me" or "I am 60% sure" are used as factors in weighting judgments. Such a policy is beneficial to the extent that confidence and accuracy are correlated (Yaniv, 1997). Finally, an expert's past performance serves as a cue to his or her accuracy. In studies in which the same experts give multiple opinions, participants form impressions about the accuracy of each expert and adjust their weights accordingly. Trust in experts is fragile, being "hard to gain, easy to lose," because negative experiences with a source have proportionally greater influence than positive ones (Yaniv & Kleinberger, 2000).

Ignoring Dissenters' Opinions

Certain configurations of opinions present particularly sharp dilemmas as to the appropriate weighting policy. Suppose that three out of four reviewers of a research proposal agree closely (consensus), but the fourth differs widely (dissension). A decision maker attempting to aggregate these opinions might rationalize the disagreement. Indeed, the need to form and maintain consonance, or harmony, is prominent in classical theories of social psychology (e.g., those of Heider and Festinger).

One mental process used to maintain consonance amounts simply to ignoring the dissonant pieces of information. Indeed, early studies of information integration and studies of judgments formed on the basis of numerical inputs of judgment (Slovic, 1966) have shown that people discount inconsistent inputs. Similarly, when intuitively combining a sample of opinions, people discount or completely ignore dissenters and assign greater weight to consensus opinions (Yaniv, 1997). Also, a dissenter's impact on a group's final decision declines as the discrepancy from the consensus increases (Davis, 1996).

On the one hand, decision makers who disregard divergent opinions could be ignoring good data because a dissenting estimate is not necessarily wrong. In general, the tendency to resolve inconsistencies by ignoring outlying views could reduce the quality of decision making. On the other hand, a policy of discounting outlying opinions might be justified if they tend to be wrong more often than consensus opinions. Certain structural aspects of the task might indicate when an outlying opinion is likely to be wrong. For example, suppose the distribution of opinions (in the population) is bell-shaped and thick-tailed. This implies that the prevalence of outlier opinions is larger than would be expected under a standard bell-shaped (normal) distribution. In such cases (assuming the bias is zero or small), an extreme opinion in a set is particularly likely to be wrong (see, e.g., DeGroot, 1986, for a discussion of the advantage of excluding outliers in estimating the center of a thick-tailed distribution). Therefore, discounting dissenters is useful if one suspects that the distribution of opinions is thick-tailed, a situation not uncommon in behavioral studies (Yaniv, 1997).

Discounting dissenters might also be justified in scenarios in which one suspects exaggeration or manipulation. For example, in certain sports competitions, such as diving and gymnastics meets, performance is evaluated by several judges whose evaluations are then combined. Suppose a judge develops a liking for a certain performer and thus, consciously or unconsciously, produces an extreme, exaggerated evaluation that could unduly affect the aggregate opinion. A common practice in combining evaluations in such competitions involves dropping the most extreme evaluations (e.g., one on each end) and averaging the middle ones. Enacting a policy that discounts extreme judgments presumably dissuades judges from acting strategically and attenuates their influence if they do so (Yaniv, 1997).

Updating One's Own Opinion: Self Versus Other

Combining one's own opinion and an advisor's opinion is a special case that requires a separate discussion. Suppose you are responsible for hiring someone to fill a job, and you initially had a strongly favorable opinion about a candidate but are told that a colleague of yours has a lukewarm opinion of the same candidate. How might you revise your opinion in light of this conflict between your own and the other opinion? You could completely ignore the other opinion, make some adjustment of your own opinion toward the other, or completely adhere to the other opinion.

From a formal point of view, other things being equal, the two opinions (own and other) might be equally weighted. However, from your internal point of view, the two opinions are not on a par. Decision makers place more weight on beliefs for which they have more evidence. Because decision makers are privy to their own thoughts, but not to the reasons underlying an advisor's opinion, they place a higher weight on their own opinion than on an advisor's. Indeed, studies show that other things being equal, people discount others' opinions and prefer their own, with the weights split roughly 70% on self and 30% on other; this balance changes when differences in ability or knowledge between self and other are made salient (Harvey & Fischer, 1997; Yaniv, 2004; Yaniv & Kleinberger, 2000). That individuals stick closely to their initial opinions is reminiscent of findings regarding attitude change—people favor their prior opinions even in the presence of contradictory evidence. But, despite the tendency to prefer one's own opinion over another person's opinion and the difficulty of assigning optimal weights to own versus other opinions, the benefit of utilizing others' estimates is appreciable. In one study (Yaniv, 2004), respondents made initial estimates of the dates of historical events and final estimates after seeing other respondents' estimates, selected at random from a pool. Using just one other opinion reduced judgment errors by about 20%.

CONCLUDING COMMENTS

Students of reasoning, judgment, and decision making have traditionally underscored the importance of generating alternatives to one's current thoughts. Other people's opinions direct decision makers to additional alternatives or unintended consequences, as these opinions may provide a different framing of a problem, an alternative explanation, or disconfirming information. Soliciting opinions is therefore an adaptive process that helps improve decisions by compensating for a pervasive weakness of human thinking.

Two theoretical issues deserve attention. First, the view of opinions as alternatives is pertinent to opinions expressed in either numerical or verbal form. Although I have focused here on combining quantitative opinions, similar psychological processes might apply to verbal opinions (advice). Surprisingly, the use of such advice in decision making has received little attention. Future research needs to consider how qualitative advice is elicited and used best.

Second, opinions about matters of fact (estimates or forecasts) differ from opinions about matters of taste (evaluations or attitudes). Theories about the benefit accrued from combining opinions about matters of fact are well developed. In contrast, simple aggregation of tastes (e.g., opinions about resorts or about types of music) for the purpose of individual decision making raises conceptual difficulties, because people are entitled to their different tastes. Nevertheless, other people's opinions about matters of taste could be used advantageously and constructively, challenging the decision maker's established preferences and inducing him or her to consider alternatives. Conceptual and empirical work is needed to clarify these issues.

Recommended Reading

Armstrong, J.S. (2001). Combining forecasts. In J.S. Armstrong (Ed.), *Principles of forecasting: A handbook for researchers and practitioners* (pp. 417–439). Norwell, MA: Kluwer.

Clemen, R.T. (1989). Combining forecasts: A review and annotated bibliography. *International Journal of Forecasting*, 5, 559–583.

Hill, G.W. (1982). Group versus individual performance: Are N+1 heads better than one? *Psychological Bulletin*, *91*, 517–539.

Acknowledgments—This research was supported by Grant No. 822 from the Israel Science Foundation.

References

Birnbaum, M.H., & Stegner, S.E. (1979). Source credibility in social judgment: Bias, expertise, and the judge's point of view. *Journal of Personality and Social Psychology*, 37, 48–74.

Davis, J.H. (1996). Group decision making and quantitative judgments: A consensus model. In E. Witte & J.H. Davis (Eds.), *Understanding group behavior: Consensual action by small groups* (pp. 35–59). Hillsdale, NJ: Erlbaum.

DeGroot, M.H. (1986). *Probability and statistics* (2nd ed.). Reading, MA: Addison-Wesley.

Einhorn, H.J., Hogarth, R.M., & Klempner, E. (1977). Quality of group judgment. *Psychological Bulletin, 84*, 158–172.

Harvey, N., & Fischer, I. (1997). Taking advice: Accepting help, improving judgment and sharing responsibility. *Organizational Behavior and Human Decision Processes, 70*, 117–133.

Hogarth, R.M. (1978). A note on aggregating opinions. *Organizational Behavior and Human Performance, 21*, 40–46.

Johnson, T.R., Budescu, D.V., & Wallsten, T.S. (2001). Averaging probability judgments: Monte Carlo analyses of asymptotic diagnostic value. *Journal of Behavioral Decision Making, 14*, 123–140.

Slovic, P. (1966). Cue-consistency and cue-utilization in judgment. *The American Journal of Psychology, 79*, 427–434.

Sniezek, J.A., & Van Swol, L.M. (2001). Trust, confidence, and expertise in a Judge-Advisor System. *Organizational Behavior and Human Decision Processes, 84*, 288–307.

Soll, J.B. (1999). Intuitive theories of information: Beliefs about the value of redundancy. *Cognitive Psychology, 38*, 317–346.

Sorkin, R.D., Hayes, C.J., & West, R. (2001). Signal detection analysis of group decision making. *Psychological Review, 108*, 183–203.

Yaniv, I. (1997). Weighting and trimming: Heuristics for aggregating judgments under uncertainty. *Organizational Behavior and Human Decision Processes, 69*, 237–249.

Yaniv, I. (2004). Receiving other people's advice: Influence and benefit. *Organizational Behavior and Human Decision Processes, 93*, 1–13.

Yaniv, I., & Kleinberger, E. (2000). Advice taking in decision making: Egocentric discounting and reputation formation. *Organizational Behavior and Human Decision Processes, 83*, 260–281.

Zarnowitz, V. (1984). The accuracy of individual and group forecasts from business and outlook surveys. *Journal of Forecasting, 3*, 11–26.

Critical Thinking Questions

1. People tend to rely on their own opinions more than the opinions of others. Why does that happen? Could it ever be rational?

2. In general it is better to rely on more than fewer opinions. How much of the benefit is due to statistical principles and how much to psychology?

3. Suppose you were in charge of evaluating an advertising campaign for a new type of product. You want to select several people from your company to figure out whether it will be successful. Who do you pick? A bunch of people who have worked together many times before and who get along well? People from different previous teams who have not worked together before and don't personally like each other? What are the cognitive and social trade-offs to each of these selections?

   (How and why would your personnel selection for this task differ from your personnel selection if you were in charge of designing an advertising campaign? See Brown in Part 2.)

Language

Language is often characterized as the pinnacle of human cognition and, indeed, understanding its cognitive basis has proven formidable. Language presents two deep problems: how it is acquired, and how it is used once it is learned. In addition to these two problems, language entails multiple levels of analysis, from individual sound units (phonemes), to words, to sentences, to stories or texts. In this section we offer a sample of each type of work.

In "Statistical Language Learning: Mechanisms and Constraints," Saffran offers a solution to a problem that, at first blush, appears virtually impossible to solve. Our perception of others' speech is that there are short pauses or breaks between the words. In fact, there are not such breaks. Speech comes to us in a relatively continuous stream, and when there are brief pauses they might well be in the middle of a word, not between words. We perceive speech as coming to us in distinct words because we are able to segment the continuous stream of sounds into discrete words. Consider, then, a baby trying to learn language. When a parent coos "Do you want the purple pacifier?" the baby hears "Doyouwantthepurplepacifier?" As far as the baby is concerned, "purple" might be a word, but so might "pur" or "plepa" or any other combination of adjacent sounds. Saffran suggests that humans are sensitive to the statistical probability of different sounds appearing next to one another, and that we use this information to segment the continuous stream of sounds into coherent words.

Once one can reliably segment words in spoken language, the challenges of language learning are not over. Estimates vary, but most college students know about sixty or seventy thousand words, most of which they encountered in print. Word learning must therefore occur at a remarkable rate, and most of it must be incidental, meaning that it's not the product of studying vocabulary in school (which researchers estimate adds only about 400 words a year). How are all of these other words learned? Landauer describes his solution in "Learning and Representing Verbal Meaning: The Latent Semantic Analysis Theory." He suggests that we must maintain a mental database that locates word meanings in a multidimensional space, similar to the way locations can be placed on a two-dimensional map. And just as familiar landmarks can help you localize an unfamiliar town without being explicitly told its location, knowledge of the surrounding words and concepts in a paragraph may help you determine the meaning of an unfamiliar word.

From language learning we turn to the use of language by experienced speakers. Obviously, knowledge of word meaning is not sufficient to ensure comprehension because word order is crucial to meaning. For example, "Sue wished she hadn't sweated" and "She wished Sue hadn't sweated" differ only in that two words have switched positions, but their meanings are altogether different. Traditional theories of sentence comprehension have assumed that the mind treats sentences in a rather algorithmic fashion, meaning that words are evaluated for the potential part of speech they might play in the sentence, and the best match is determined by a set of rules for syntactic assignment. In "Good-enough Representations in Language Comprehension," Ferreira, Bailey, and Ferraro review recent research that offers a different interpretation. Sentence syntax is one source of information to help derive meaning from a sentence, but there are others, such as context. People may only process syntax to the point that they are satisfied that they have derived a meaning that is sensible. Surprisingly, when it becomes clear that this meaning is incorrect, the old meaning seems to persist, along with the new, correct meaning.

Even if we fully understood how sentences are comprehended, we still would not have a complete account of language. Research shows that when we read a story, for example, we keep track of what is happening in the story. This representation is called a situation model. For example, if we are told that a character is in her apartment, we note that fact, and if we are later told that she goes down a stairway to the basement of the building, we update the situation model to reflect her new location. In "Situation Models: The Mental Leap into Imagined Worlds," Zwaan reviews the basic components of a story that are maintained in the situation model, and also presents evidence that the reader sees him or herself as being *in* the narrative; when our hero goes downstairs to the basement, we go with her.

The four articles in this section offer the latest perspectives on the cognitive processes that support language; however, they also offer a perspective on a large-scale issue of enduring interest: is language innate or learned? Most any researcher would argue that the answer must be "both"; we learn language, but are biologically prepared to do so, and the relative importance of learning and innateness varies with different aspects of language. Nevertheless, it's worth noting that two of the articles in this section—Saffran and Landauer—show the power of relatively simple learning processes that focus on key information in the environment, and accrue learning over the long term.

Statistical Language Learning: Mechanisms and Constraints

Jenny R. Saffran[1]
Department of Psychology and Waisman Center, University of Wisconsin-Madison, Madison, Wisconsin

Abstract

What types of mechanisms underlie the acquisition of human language? Recent evidence suggests that learners, including infants, can use statistical properties of linguistic input to discover structure, including sound patterns, words, and the beginnings of grammar. These abilities appear to be both powerful and constrained, such that some statistical patterns are more readily detected and used than others. Implications for the structure of human languages are discussed.

Keywords

language acquisition; statistical learning; infants

Imagine that you are faced with the following challenge: You must discover the underlying structure of an immense system that contains tens of thousands of pieces, all generated by combining a small set of elements in various ways. These pieces, in turn, can be combined in an infinite number of ways, although only a subset of those combinations is actually correct. However, the subset that is correct is itself infinite. Somehow you must rapidly figure out the structure of this system so that you can use it appropriately early in your childhood.

This system, of course, is human language. The elements are the sounds of language, and the larger pieces are the words, which in turn combine to form sentences. Given the richness and complexity of language, it seems improbable that children could ever discern its structure. The process of acquiring such a system is likely to be nearly as complex as the system itself, so it is not surprising that the mechanisms underlying language acquisition are a matter of long-standing debate. One of the central focuses of this debate concerns the innate and environmental contributions to the language-acquisition process, and the degree to which these components draw on information and abilities that are also relevant to other domains of learning.

In particular, there is a fundamental tension between theories of language acquisition in which learning plays a central role and theories in which learning is relegated to the sidelines. A strength of learning-oriented theories is that they exploit the growing wealth of evidence suggesting that young humans possess powerful learning mechanisms. For example, infants can rapidly capitalize on the statistical properties of their language environments, including the distributions of sounds in words and the orders of word types in sentences, to discover important components of language structure. Infants can track such statistics, for example, to discover speech categories (e.g., native-language consonants; see, e.g., Maye, Werker, & Gerken, 2002), word boundaries (e.g., Saffran, Aslin, & Newport, 1996), and rudimentary syntax (e.g., Gomez & Gerken, 1999; Saffran & Wilson, 2003).

However, theories of language acquisition in which learning plays a central role are vulnerable to a number of criticisms. One of the most important arguments against learning-oriented theories is that such accounts seem at odds with one of the central observations about human languages. The linguistic systems of the world, despite surface differences, share deep similarities, and vary in nonarbitrary ways. Theories of language acquisition that focus primarily on preexisting knowledge of language do provide an elegant explanation for cross-linguistic similarities. Such theories, which are exemplified by the seminal work of Noam Chomsky, suggest that linguistic universals are prespecified in the child's linguistic endowment, and do not require learning. Such accounts generate predictions about the types of patterns that should be observed cross-linguistically, and lead to important claims regarding the evolution of a language capacity that includes innate knowledge of this kind (e.g., Pinker & Bloom, 1990).

Can learning-oriented theories also account for the existence of language universals? The answer to this question is the object of current research. The *constrained statistical learning framework* suggests that learning is central to language acquisition, and that the specific nature of language learning explains similarities across languages. The crucial point is that learning is constrained; learners are not open-minded, and calculate some statistics more readily than others. Of particular interest are those constraints on learning that correspond to cross-linguistic similarities (e.g., Newport & Aslin, 2000). According to this framework, the similarities across languages are indeed nonaccidental, as suggested by the Chomskian framework—but they are not the result of innate linguistic knowledge. Instead, human languages have been shaped by human learning mechanisms (along with constraints on human perception, processing, and speech production), and aspects of language that enhance learnability are more likely to persist in linguistic structure than those that do not. Thus, according to this view, the similarities across languages are not due to innate knowledge, as is traditionally claimed, but rather are the result of constraints on learning. Further, if human languages were (and continue to be) shaped by constraints on human learning mechanisms, it seems likely that these mechanisms and their constraints were not tailored solely for language acquisition. Instead, learning in nonlinguistic domains should be similarly constrained, as seems to be the case.

A better understanding of these constraints may lead to new connections between theories focused on nature and theories focused on nurture. Constrained learning mechanisms require both particular experiences to drive learning and preexisting structures to capture and manipulate those experiences.

LEARNING THE SOUNDS OF WORDS

In order to investigate the nature of infants' learning mechanisms, my colleagues and I began by studying an aspect of language that we knew must certainly be learned: word segmentation, or the boundaries between words in fluent speech. This is a challenging problem for infants acquiring their first language, for speakers do not mark word boundaries with pauses, as shown in Figure 1. Instead, infants must determine where one word ends and the next begins without access

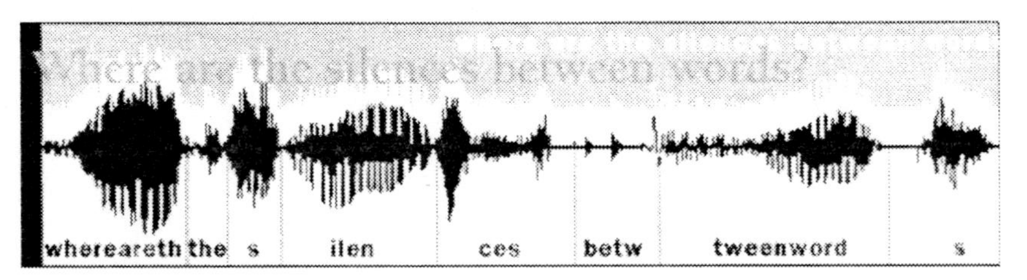

Fig. 1. A speech waveform of the sentence "Where are the silences between words?" The height of the bars indicates loudness, and the *x*-axis is time. This example illustrates the lack of consistent silences between word boundaries in fluent speech. The vertical gray lines represent quiet points in the speech stream, some of which do not correspond to word boundaries. Some sounds are represented twice in the transcription below the waveform because of their continued persistence over time.

to obvious acoustic cues. This process requires learning because children cannot innately know that, for example, *pretty* and *baby* are words, whereas *tyba* (spanning the boundary between *pretty* and *baby*) is not.

One source of information that may contribute to the discovery of word boundaries is the statistical structure of the language in the infant's environment. In English, the syllable *pre* precedes a small set of syllables, including *ty*, *tend*, and *cedes*; in the stream of speech, the probability that *pre* is followed by *ty* is thus quite high (roughly 80% in speech to young infants). However, because the syllable *ty* occurs word finally, it can be followed by any syllable that can begin an English word. Thus, the probability that *ty* is followed by *ba*, as in *pretty baby*, is extremely low (roughly 0.03% in speech to young infants). This difference in sequential probabilities is a clue that *pretty* is a word, and *tyba* is not. More generally, given the statistical properties of the input language, the ability to track sequential probabilities would be an extremely useful tool for infant learners.

To explore whether humans can use statistical learning to segment words, we exposed adults, first graders, and 8-month-olds to spoken nonsense languages in which the only cues to word boundaries were the statistical properties of the syllable sequences (e.g., Saffran et al., 1996). Listeners briefly heard a continuous sequence of syllables containing multisyllabic words from one of the languages (e.g., *golabupabikututibubabupugolabubabupu* . . .). We then tested our participants to determine whether they could discriminate the words from the language from sequences spanning word boundaries. For example, we compared performance on words like *golabu* and *pabiku* with performance on sequences like *bupabi*, which spanned the boundary between words. To succeed at this task, listeners would have had to track the statistical properties of the input. Our results confirmed that human learners, including infants, can indeed use statistics to find word boundaries. Moreover, this ability is not confined to humans: Cotton-top tamarins, a New World monkey species, can also track statistics to discover word boundaries (Hauser, Newport, & Aslin, 2001).

One question immediately raised by these results is the degree to which statistical learning is limited to language-like stimuli. A growing body of results suggests that sequential statistical learning is quite general. For example, infants can track sequences of tones, discovering "tone-word boundaries" via statistical cues (e.g., Saffran, Johnson, Aslin, & Newport, 1999), and can learn statistically defined visual patterns (e.g., Fiser & Aslin, 2002; Kirkham, Slemmer, & Johnson, 2002); work in progress is extending these results to the domain of events in human action sequences.

Given that the ability to discover units via their statistical coherence is not confined to language (or to humans), one might wonder whether the statistical learning results actually pertain to language at all. That is, do infants actually use statistical learning mechanisms in real-world language acquisition? One way to address this question is to ask what infants are actually learning in our segmentation task. Are they learning statistics? Or are they using statistics to learn language? Our results suggest that when infants being raised in English-speaking environments have segmented the sound strings, they treat these nonsensical patterns as English words (Saffran, 2001b). Statistical language learning in the laboratory thus appears to be integrated with other aspects of language acquisition. Related results suggest that 12-month-olds can first segment novel words and then discover syntactic regularities relating the new words—all within the same set of input. This would not be possible if the infants formed mental representations only of the sequential probabilities relating individual syllables, and no word-level representations (Saffran & Wilson, 2003). These findings point to a constraint on statistical language learning: The mental representations produced by this process are not just sets of syllables linked by statistics, but new units that are available to serve as the input to subsequent learning processes.

Similarly, it is possible to examine constraints on learning that might affect the acquisition of the sound structure of human languages. The types of sound patterns that infants learn most readily may be more prevalent in languages than are sound patterns that are not learnable by infants. We tested this hypothesis by asking whether infants find some phonotactic regularities (restrictions on where particular sounds can occur; e.g., /fs/ can occur at the end, but not the beginning, of syllables in English) easier to acquire than others (Saffran & Thiessen, 2003). The results suggest that infants readily acquire novel regularities that are consistent with the types of patterns found in the world's languages, but fail to learn regularities that are inconsistent with natural language structure. For example, infants rapidly learn new phonotactic regularities involving generalizations across sounds that share a phonetic feature, while failing to learn regularities that disregard such features. Thus, it is easier for infants to learn a set of patterns that group together /p/, /t/, and /k/, which are all voiceless, and that group together /b/, /d/, and /g/, which are all voiced, than to learn a pattern that groups together /d/, /p/, and /k/, but does not apply to /t/.[2] Studies of this sort may provide explanations for why languages show the types of sound patterning that they do; sound structures that are hard for infants to learn may be unlikely to recur across the languages of the world.

STATISTICAL LEARNING AND SYNTAX

Issues about learning versus innate knowledge are most prominent in the area of syntax. How could learning-oriented theories account for the acquisition of abstract structure (e.g., phrase boundaries) not obviously mirrored in the surface statistics of the input? Unlike accounts centered on innate linguistic knowledge, most learning-oriented theories do not provide a transparent explanation for the ubiquity of particular structures cross-linguistically. One approach to these issues is to ask whether some nearly universal structural aspects of human languages may result from constraints on human learning (e.g., Morgan, Meier, & Newport, 1987). To test this hypothesis, we asked whether one such aspect of syntax, phrase structure (groupings of types of words together into subunits, such as noun phrases and verb phrases), results from a constraint on learning: Do humans learn sequential structures better when they are organized into subunits such as phrases than when they are not? We identified a statistical cue to phrasal units, predictive dependencies (e.g., the presence of a word like *the* or *a* predicts a noun somewhere downstream; the presence of a preposition predicts a noun phrase somewhere downstream), and determined that learners can use this kind of cue to locate phrase boundaries (Saffran, 2001a).

In a direct test of the theory that predictive dependencies enhance learnability, we compared the acquisition of two nonsense languages, one with predictive dependencies as a cue to phrase structure, and one lacking predictive dependencies (e.g., words like *the* could occur either with or without a noun, and a noun could occur either with or without words like *the*; neither type of word predicted the presence of the other). We found better language learning in listeners exposed to languages containing predictive dependencies than in listeners exposed to languages lacking predictive dependencies (Saffran, 2002). Interestingly, the same constraint on learning emerged in tasks using nonlinguistic materials (e.g., computer alert sounds and simultaneously presented shape arrays). These results support the claim that learning mechanisms not specifically designed for language learning may have shaped the structure of human languages.

DIRECTIONS FOR FUTURE RESEARCH

Results to date demonstrate that human language learners possess powerful statistical learning capacities. These mechanisms are constrained at multiple levels; there are limits on what information serves as input, which computations are performed over that input, and the structure of the representations that emerge as output. To more fully understand the contribution of statistical learning to language acquisition, it is necessary to assess the degree to which statistical learning provides explanatory power given the complexities of the acquisition process.

For example, how does statistical learning interact with other aspects of language acquisition? One way we are addressing this question is by investigating how infants weight statistical cues relative to other cues to word segmentation early in life. The results of such studies provide an important window into the ways in which statistical learning may help infant learners to determine the relevance of the many cues inherent in language input. Similarly, we are studying how statistics meet up with meaning in the world (e.g., are statistically defined "words" easier to learn as labels for novel objects than sound sequences spanning word boundaries?), and how infants in bilingual environments cope with multiple sets of statistics. Studying the intersection between statistical learning and the rest of language learning may provide new insights into how various nonstatistical aspects of language are acquired. Moreover, a clearer picture of the learning mechanisms used successfully by typical language learners may increase researchers' understanding of the types of processes that go awry when children do not acquire language as readily as their peers.

It is also critical to determine which statistics are available to young learners and whether those statistics are actually relevant to natural language structure. Researchers do not agree on the role that statistical learning should play in acquisition theories. For example, they disagree about when learning is best described as statistically based as opposed to rule based (i.e., utilizing mechanisms that operate over algebraic variables to discover abstract knowledge), and about whether learning can still be considered statistical when the input to learning is abstract. Debates regarding the proper place for statistical learning in theories of language acquisition cannot be resolved in advance of the data. For example, although one can distinguish between statistical versus rule-based learning mechanisms, and statistical versus rule-based knowledge, the data are not yet available to determine whether statistical learning itself renders rule-based knowledge structures, and whether abstract knowledge can be probabilistic. Significant empirical advances will be required to disentangle these and other competing theoretical distinctions.

Finally, cross-species investigations may be particularly informative with respect to the relationship between statistical learning and human language. Current research is identifying species differences in the deployment of statistical learning mechanisms (e.g., Newport & Aslin, 2000). To the extent that nonhumans and humans track different statistics, or track statistics over different perceptual units, learning mechanisms that do not initially appear to be human-specific may actually render human-specific outcomes. Alternatively, the overlap between the learning mechanisms available across species may suggest that differences in statistical learning cannot account for cross-species differences in language-learning capacities.

CONCLUSION

It is clear that human language is a system of mind-boggling complexity. At the same time, the use of statistical cues may help learners to discover some of the patterns lurking in language input. To what extent might the kinds of statistical patterns accessible to human learners help in disentangling the complexities of this system? Although the answer to this question remains unknown, it is possible that a combination of inherent constraints on the types of patterns acquired by learners, and the use of output from one level of learning as input to the next, may help to explain why something so complex is mastered readily by the human mind. Human learning mechanisms may themselves have played a prominent role in shaping the structure of human languages.

Recommended Reading

Gómez, R.L., & Gerken, L.A. (2000). Infant artificial language learning and language acquisition. *Trends in Cognitive Sciences, 4*, 178–186.

Hauser, M.D., Chomsky, N., & Fitch, W.T. (2002). The faculty of language: What is it, who has it, and how did it evolve? *Science*, *298*, 1569–1579.
Pena, M., Bonatti, L.L., Nespor, M., & Mehler, J. (2002). Signal-driven computations in speech processing. *Science*, *298*, 604–607.
Seidenberg, M.S., MacDonald, M.C., & Saffran, J.R. (2002). Does grammar start where statistics stop? *Science*, *298*, 553–554.

Acknowledgments—The preparation of this manuscript was supported by grants from the National Institutes of Health (HD37466) and National Science Foundation (BCS-9983630). I thank Martha Alibali, Erin McMullen, Seth Pollak, Erik Thiessen, and Kim Zinski for comments on a previous version of this manuscript.

Notes

1. Address correspondence to Jenny R. Saffran, Department of Psychology, University of Wisconsin-Madison, Madison, WI 53706; e-mail: jsaffran@wisc.edu.

2. Voicing refers to the timing of vibration of the vocal cords. Compared with voiceless consonants, voiced consonants have a shorter lag time between the initial noise burst of the consonant and the subsequent vocal cord vibrations.

References

Fiser, J., & Aslin, R.N. (2002). Statistical learning of new visual feature combinations by infants. *Proceedings of the National Academy of Sciences, USA*, *99*, 15822–15826.
Gomez, R.L., & Gerken, L. (1999). Artificial grammar learning by 1-year-olds leads to specific and abstract knowledge. *Cognition*, *70*, 109–135.
Hauser, M., Newport, E.L., & Aslin, R.N. (2001). Segmentation of the speech stream in a nonhuman primate: Statistical learning in cotton-top tamarins. *Cognition*, *78*, B41–B52.
Kirkham, N.Z., Slemmer, J.A., & Johnson, S.P. (2002). Visual statistical learning in infancy: Evidence of a domain general learning mechanism. *Cognition*, *83*, B35–B42.
Maye, J., Werker, J.F., & Gerken, L. (2002). Infant sensitivity to distributional information can affect phonetic discrimination. *Cognition*, *82*, B101–B111.
Morgan, J.L., Meier, R.P., & Newport, E.L. (1987). Structural packaging in the input to language learning: Contributions of intonational and morphological marking of phrases to the acquisition of language. *Cognitive Psychology*, *19*, 498–550.
Newport, E.L., & Aslin, R.N. (2000). Innately constrained learning: Blending old and new approaches to language acquisition. In S.C. Howell, S.A. Fish, & T. Keith-Lucas (Eds.), *Proceedings of the 24th Boston University Conference on Language Development* (pp. 1–21). Somerville, MA: Cascadilla Press.
Pinker, S., & Bloom, P. (1990). Natural language and natural selection. *Behavioral and Brain Sciences*, *13*, 707–784.
Saffran, J.R. (2001a). The use of predictive dependencies in language learning. *Journal of Memory and Language*, *44*, 493–515.
Saffran, J.R. (2001b). Words in a sea of sounds: The output of statistical learning. *Cognition*, *81*, 149–169.
Saffran, J.R. (2002). Constraints on statistical language learning. *Journal of Memory and Language*, *47*, 172–196.
Saffran, J.R., Aslin, R.N., & Newport, E.L. (1996). Statistical learning by 8-month-old infants. *Science*, *274*, 1926–1928.
Saffran, J.R., Johnson, E.K., Aslin, R.N., & Newport, E.L. (1999). Statistical learning of tone sequences by human infants and adults. *Cognition*, *70*, 27–52.
Saffran, J.R., & Thiessen, E.D. (2003). Pattern induction by infant language learners. *Developmental Psychology*, *39*, 484–494.
Saffran, J.R., & Wilson, D.P. (2003). From syllables to syntax: Multi-level statistical learning by 12month-old infants. *Infancy*, *4*, 273–284.

Critical Thinking Questions

1. Foreign languages often sound as if they are spoken rapidly, and without pauses between words. Explain why that is so, given Saffran's explanation of statistical language learning.

2. Saffran suggests that the statistical properties of language help us to learn the boundaries of words, and also help us to learn syntax. But we also have knowledge of higher-level structures in language. For example, we know the basic structure that a story follows, commonly called a story grammar: a story has a main character with a goal, but there are obstacles to achieving the goal, and so forth. Could the statistical learning framework be extended to account for high-level language knowledge like that represented in story grammars?

3. The fact that tamarins are sensitive to the statistical properties of linguistic input might make us think that this sensitivity is a general learning ability of primates. In turn, that might make us think that this learning ability is not special to language, but rather, we are, in general, good at learning the statistical properties of sequences that we experience. From your personal experience, do you think that's true?

Learning and Representing Verbal Meaning: The Latent Semantic Analysis Theory

Thomas K. Landauer[1]
Department of Psychology, University of Colorado, Boulder, Colorado

Abstract

Latent semantic analysis (LSA) is a theory of how word meaning—and possibly other knowledge—is derived from statistics of experience, and of how passage meaning is represented by combinations of words. Given a large and representative sample of text, LSA combines the way thousands of words are used in thousands of contexts to map a point for each into a common semantic space. LSA goes beyond pairwise co-occurrence or correlation to find latent dimensions of meaning that best relate every word and passage to every other. After learning from comparable bodies of text, LSA has scored almost as well as humans on vocabulary and subject-matter tests, accurately simulated many aspects of human judgment and behavior based on verbal meaning, and been successfully applied to measure the coherence and conceptual content of text. The surprising success of LSA has implications for the nature of generalization and language.

Keywords

latent semantic analysis; latent semantic indexing; LSA; learning; meaning; lexicon; knowledge; machine learning; simulation

By age 18, you knew the meaning of more than 50,000 words that you had met only in print. How did you do that? My colleagues and I think that we may have cracked this and some other persistent mysteries of verbal meaning. We have been exploring a mathematical computer model and corresponding psychological learning theory called *Latent Semantic Analysis* (LSA). Although far from perfect or complete as a theory of meaning and language, LSA accurately simulates many aspects of human understanding of word and passage meaning and can effectively replace human text comprehension in several educational applications. Among other things, it mimics the rate at which schoolchildren learn recognition vocabulary from text, makes humanlike assessments of semantic relationships between words, passes college multiple-choice exams after "reading" a textbook, and makes it possible to automatically assess the content of factual essays as reliably as expert humans.

THE LATENT SEMANTIC ANALYSIS THEORY

The formal LSA model relies on sophisticated mathematical and computer methods—ones we think may mirror what the brain does (Landauer & Dumais, 1997; Landauer, Foltz, & Laham, 1998). Although a technical discussion of these methods is beyond the scope of this review, a nontechnical description can show what LSA assumes and how it works.

LSA expresses a venerable idea about word meaning, that words occupy positions in a *semantic space* and their meaning is the relation of each to all. Since the arrival of computers, the idea has often taken the form of programs in which words are linked by labels such as *part of*, or through common features like *is living*. Unfortunately, because knowledgeable humans supply the links and labels, these programs beg the question in which we are most interested: how people acquire meaning from experience in the first place.

The psychological theory underlying LSA assumes that people start by associating perceptual objects and experiences, including words, that are met near each other in time. Doing this helps people predict the world in advance and deal with it better as a result. But human cognition (at least) goes far beyond piecewise association. It somehow takes all the billions of local contiguity relations and fits them together into an overall map, a semantic space that represents how each object, event, or word is related to each other.

LSA as mathematical computer model also constructs a semantic space. It gets its experience by being fed a large body of electronic text. It starts by using a computer version of associative learning to establish a link between each unique word type (e.g., the word *psychology* wherever it appears) and every paragraph in which it appears (say, this one). This step is essential, and the particular way it is done matters. However, the real power of LSA comes from a succeeding process, one that combines these billions of little links to form a common semantic space in which each word and any passage in the language has its own place.

LSA does this in a way that is analogous to how cartographers once mapped the Earth. They started with rough estimates of point-to-point distances based on sightings from hilltops, camel-travel days, sailing times, or the likelihood of anyone getting from there to here. They put all of this piecewise data together by placing the points—towns, river junctions, islands, headlands, mythic places—onto a single picture in a way that preserved all the measurements as well as possible. If successful., they not only improved the distance estimates and produced a pleasing graphic, but got an immense added benefit in the ability to read off the infinite number of point-to-point distances that had never been measured. Such maps let people understand their world in an entirely new and better way.

This kind of mapping works because of a simple fact of geometry. A structure of points in which each is connected to many others in properly interlocking triangles is rigid, so any missing paths are strictly defined. However, this works well only if you have assumed the correct shape—the right dimensionality—of the surface that the points are mapped onto. For example, mapping the spherical surface of the Earth onto a flat piece of paper creates major distortions; a globe does much better.

LSA assumes that the mind or brain applies essentially the same method to mapping the meaning of words (and other experiences) into semantic space. It starts with rough estimates of closeness in the form of local temporal associations, then finds a way to fit them together. For LSA's initial estimate of the closeness of two words, it observes how often they occur in the same meaningful contexts (such as sentences or paragraphs) relative to how often they occur in different contexts, and computes a correlation index that approximates Pavlovian conditioning. However, this is just the first step; the correlations themselves are not the answer. Instead, LSA fits all the separate relations into a common space in a manner—and dimensionality—that distorts them as little as possible. It is this second step that constitutes its understanding of meaning.

In the case of physical mapping, two towns may be quite close, but because they are on different sides of a mountain, the distance between them may never have been measured. Nevertheless, it is easily determined from a map. Similarly, two words may be quite alike in meaning but rarely used in the same context, for example, because they are synonyms. However, when LSA places them in semantic space, it can bring them close together (or, if the data so imply, spread them out).

In applying LSA as a simulation model, using the right number of dimensions can make a big difference, and the best number is usually between 100 and 1,000, not the two or three of physical maps. My colleagues and I believe that the high dimensionality needed by LSA is a product of the way the brain is structured combined with the statistical structure of experience.

SIMULATIONS

An invaluable feature of LSA—a result of advances in computer power and the availability of large amounts of electronic text—is that we can often have the model learn from almost the same sources, both in size and in content, as a human does. We can then test LSA in some of the same ways that humans are tested and have it perform some of the same meaning-based tasks that humans perform. We believe that central human cognitive abilities often depend on immense amounts of experience, and that theories that cannot be applied to comparable data may be fundamentally wrong, or at least unprovable. One reason that we have applied LSA to practical problems is to assess how well it captures everything that is important.

The best example of this strategy comes from simulating schoolchildren's learning of vocabulary from reading. Of the roughly 400,000 words that a typical seventh grader could have encountered in print but nowhere else, she knows the meaning of 10 today that she did not know yesterday. Amazingly, she saw only 2 or 3 of these in the interim. Moreover, when psychologists have tried to teach word meanings explicitly, they have not come close to normal rates. Traditional theories of word learning, which are based on specific concrete experiences with individual words, cannot account for these facts. However, when a very large number of other words is known, the indirect effects assumed in LSA offer an explanation: Correctly mapping a few new relations can help to define many more.

Dumais and I had LSA learn from samples of 5 million words of natural text, comparable to the total lifetime reading of a seventh grader, then take standard multiple-choice vocabulary tests (Landauer & Dumais, 1997). From the results, we calculated that if LSA were exposed to the same reading materials as the student, it too would know about 10 new words each day. Moreover, by varying the content of the training text, we deduced that about three fourths of LSA's gain per paragraph was indirect, knowing more about words the model did not encounter because of how the ones it did encounter improved the overall semantic space.

We have also looked in more detail at LSA's knowledge of word meanings, showing, for example, that it usually represents synonym and antonym pairs, singulars and plurals, and members of the same conceptual categories as related to each other in the way people think they are. It also mimics the way that words like ball have more than one meaning, and the way that such words are disambiguated by context, simulating closely the results of classic psycholinguistic laboratory experiments.

We have also trained LSA on more focused bodies of knowledge. Figure 1 shows the results when LSA learned from an electronic version of an introductory psychology textbook, then was tested on the same multiple-choice exams used in large classes at two universities. For each test item, LSA separately represented the question and each alternative as points in its semantic space by averaging points for their contained words. It then chose the alternative most similar to the question. LSA did not do quite as well as the average student (who, unlike LSA, had also attended lectures), but did well enough to pass and showed the same pattern of better performance on easy (for students) factual than difficult conceptual questions.

APPLICATIONS

A set of educational applications of LSA provides additional, and we think important, confirmation of its ability to capture meaning. LSA represents the meaning of a passage just as it does the meaning of a word, as a single point

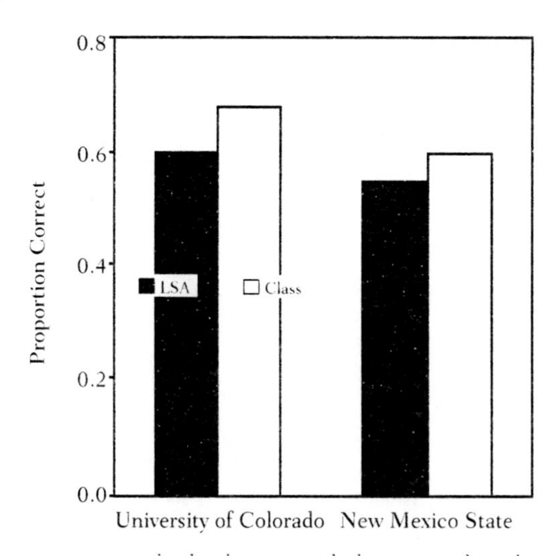

Fig. 1. Performance on a multiple-choice psychology exam by a latent semantic analysis (LSA) simulation and students at two universities. For the classes, mean scores are shown.

in its semantic space. The point for a passage is just the center, or average, of all the words it contains, independent of their order. This means that LSA ignores any meaning that could not be recovered after a passage is scrambled, a possibly serious limitation. However, the single-point representation also implies that the same meaning, as represented in LSA, can be verbally expressed by many different wordings, an important ability if it is to mimic human understanding.

In one application, we accurately predicted how well students would comprehend an expository text about the heart by having LSA compute how similar the meaning of each paragraph was to the one before it. In another, we had students first write short essays on the heart, then correctly predicted that they would learn most from a text that was a little, but not too much, more sophisticated than their essays, measuring the content of the essays, the content of the texts, and their relations with LSA alone.

In the most recent application, we have used LSA to measure the quality and quantity of conceptual content in essay answers to questions about factual subjects. By comparing the LSA representation of a student's essay to a large number of essays on the same topic that have been scored by an expert human, we can accurately estimate the score that the same expert would give a new essay. As shown in Figure 2, over some 1,200 individual essays on seven diverse topics, the correlation between LSA and a human expert was virtually identical to the correlation between one human expert and another.

We have also successfully field-tested a system that lets sixth graders write summaries of chapters, get quality scores and automatically generated commentaries, and then try again until they are ready to turn their summaries in to the teacher. We think such systems can greatly increase the number of opportunities for valuable practice in understanding and composing expository text.

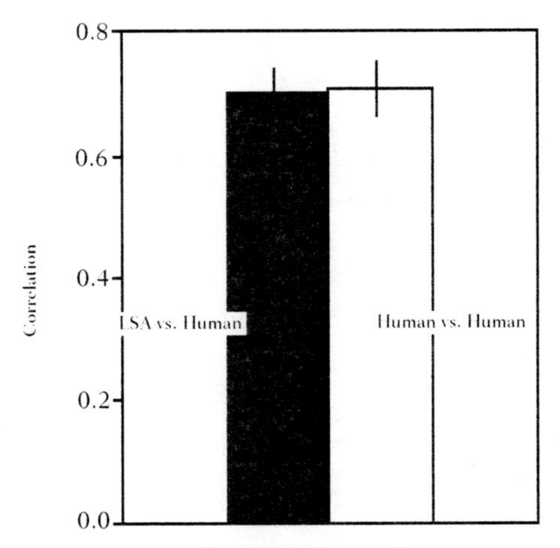

Fig. 2. Average agreement (correlation) between latent semantic analysis (LSA) and human expert scoring of 1,203 essay-exam answers on topics in biology, psychology, history, sociology, and business.

IMPLICATIONS

The rather remarkable successes of the LSA model raise a variety of interesting theoretical questions and suggest a number of potentially practical uses.

- *Generalization.* LSA induces similarity relations among words from contextual experience alone rather than from the sharing of primitive features; LSA knows nothing about what two words have in common except how they occur in the company of other words. Similar mechanisms may help to explain other similarity-based phenomena, such as object recognition.
- *Innateness and experience.* LSA's simulation successes prove that it is possible to induce much more humanlike knowledge from experience than is apparent in surface correlations. This calls into question intuitions about "the poverty of the stimulus," the notion that what people know far exceeds what they could have learned, an apparent paradox that has plagued theorists of meaning from Plato through Wittgenstein and Quine to Chomsky and Pinker.
- *Relative importance of words and syntax.* LSA captures a surprising amount of word and passage meaning without paying attention to word order within sentences and paragraphs. It may be that people, too, represent verbal meaning primarily as word combinations. Perhaps syntax largely serves other functions—such as easing encoding or decoding—and is relatively less important for storing meaning than has usually been assumed in linguistics and psychology.
- *Utility.* Because LSA can mirror human comprehension of language, there are many human cognitive tasks for which it may offer significant help. Comparisons of meaning that are too cumbersome for human processing may become routine. For example, in its oldest application, LSA improved automatic information retrieval by matching queries and documents on meaning rather than words, and in recent pilot experiments, LSA measured the conceptual overlap between 4,000 pairs of military training courses in a few minutes.

In sum, we think LSA offers exciting possibilities on many fronts. There is, of course, much work left to do. We would like to find ways to improve LSA's learning databases, to include syntax-based meaning, to make LSA generate as well as understand discourse, and to test more directly its psychological and neural reality. Meanwhile, to see what it does now, and to use its products for new research and applications, you can visit the World Wide Web: http://LSA.colorado.edu/.

Recommended Reading

Kintsch, W. (1998). *Comprehension: A paradigm for cognition.* Cambridge, England: Cambridge University Press.

Landauer, T.K., & Dumais, S.T. (1997). (See References)

Landauer, T.K., Foltz, P., & Laham, R.D. (1998). (See References)

Miller, G.A. (1996). *The science of words.* New York: W.H. Freeman.

Acknowledgments—I thank many colleagues and collaborators, especially Curt Burgess, Susan Dumais, Peter Foltz, Walter Kintsch, Darrell Laham, Bob Rehder Missy Schreiner, and Michael Wolfe. Financial support was provided by Bellcore, the McDonnell Foundation, and the Defense Advanced Research Project Administration. Stephen Ivens of Touchstone Applied Science Associates and Patrick Kylonnen of the U.S. Air Force Research Laboratory generously shared data.

Note

1. Address correspondence to Thomas K. Landauer, Department of Psychology, Campus Box 344, University of Colorado, Boulder, CO 80309-0344.

References

Landauer, T.K., & Dumais, S.T. (1997). A solution to Plato's Problem: The Latent Semantic Analysis theory of acquisition, induction and representation of knowledge. *Psychological Review, 104*, 211–240.

Landauer, T.K., Foltz, P., & Laham, R.D. (1998). An introduction to latent semantic analysis. *Discourse Processes, 25*, 259–284.

Critical Thinking Questions

1. This article presents a model that attempts to solve two of the most challenging problems in language: first, how meaning is acquired from experience, and second, how we are able to learn word meanings at such a rapid rate through childhood. How does the model solve each of these problems?
2. The model assumes that word meaning is represented in a multidimensional space. To appreciate what is meant by "space" the author draws an analogy to a map. You can locate a town by knowing its location on two dimensions: a north-south dimension and an east-west dimension. Thus a map is a two-dimensional space. Semantic space will have not two dimensions, but 100 or 1000, and they will not all be spatial, but ones that create meaning. But where do these dimensions come from? Words have meaning because they are characterized on these 100 or 1000 dimensions—but how do the dimensions have meaning?
3. The model accounts for how new words and concepts are acquired. If the model is correct, what sort of reading should you do to maximize how much you learn from what you read?

Good-Enough Representations in Language Comprehension

Fernanda Ferreira,[1] Karl G.D. Bailey, and Vittoria Ferraro
Department of Psychology and Cognitive Science Program, Michigan State University, East Lansing, Michigan

Abstract

People comprehend utterances rapidly and without conscious effort. Traditional theories assume that sentence processing is algorithmic and that meaning is derived compositionally. The language processor is believed to generate representations of the linguistic input that are complete, detailed, and accurate. However, recent findings challenge these assumptions. Investigations of the misinterpretation of both garden-path and passive sentences have yielded support for the idea that the meaning people obtain for a sentence is often not a reflection of its true content. Moreover, incorrect interpretations may persist even after syntactic reanalysis has taken place. Our good-enough approach to language comprehension holds that language processing is sometimes only partial and that semantic representations are often incomplete. Future work will elucidate the conditions under which sentence processing is simply good enough.

Keywords

language comprehension; satisficing; syntax; linguistic ambiguity

Over the past three decades, various theories of language comprehension have been developed to explain how people compose the meanings of sentences from individual words. All theories advanced to date assume that the language-processing mechanism applies a set of algorithms to access words from the lexicon, organize them into a syntactic structure through rules of grammar, and derive the meaning of the whole structure based on the meaning of its parts. Furthermore, all theories assume that this process generates complete, detailed, and accurate representations of the linguistic input.

MODELS OF SENTENCE PROCESSING

Two approaches to sentence processing that have been widely contrasted are the *garden-path model* (Ferreira & Clifton, 1986; Frazier, 1978) and the *constraint-satisfaction model* (MacDonald, Pearlmutter, & Seidenberg, 1994; Trueswell, Tanenhaus, & Garnsey, 1994). According to the garden-path account, the language processor initially computes a single syntactic analysis without consideration of context or plausibility. Once an interpretation has been chosen, other information is used to evaluate its appropriateness. For example, a person who heard, "Mary saw the man with the binoculars," would tend to understand the sentence to mean that Mary used the binoculars as an instrument. If it turned out that the man had the binoculars, the initial interpretation would be revised to be compatible with that contextual knowledge.

Constraint-satisfaction theorists, in contrast, assume that all possible syntactic analyses are computed at once on the basis of all relevant sources of information. The analysis with the greatest support is chosen over its competitors. The constraint-based approach predicts that people who hear the sentence about Mary, the man, and the binoculars will activate both interpretations and then select the one that is more appropriate in the context. Thus, the two classes of models assume radically different approaches to sentence processing: According to the garden-path model, analyses are proposed serially, and syntactic information is processed entirely separately from real-world knowledge and meaning. According to constraint-based models, analyses are proposed in parallel, and the syntactic processor communicates with any relevant information source. Nevertheless, both models incorporate the assumption that interpretations of utterances are compositionally built up from words clustered into hierarchically organized constituents.

IS THE MEANING OF A SENTENCE ALWAYS THE SUM OF ITS PARTS?

This assumption of compositionality seems eminently plausible, but results in the literature on the psychology of language call it into question. For example, people have been observed to unconsciously normalize strange sentences to make them sensible (Fillenbaum, 1974). The *Moses illusion* (Erickson & Mattson, 1981) is typically viewed as demonstrating the fallibility of memory processes, but it is also relevant to issues of language interpretation and compositionality. When asked, "How many animals of each sort did Moses put on the ark?" people tend to respond "two," instead of objecting to the presupposition behind the question. Similarly, participants often overlook the anomaly in a sentence such as "The authorities had to decide where to bury the survivors" (Barton & Sanford, 1993).

A study conducted to examine whether sentence meaning can prime individual words (i.e., activate them so that they are more accessible to the comprehension system) also demonstrates that language processing is not always compositional, and that the semantic representations that get computed are shallow and incomplete (rather than computing the structure to the fullest degree possible, the comprehension system just does enough to contend with the overall task at hand; Duffy, Henderson, & Morris, 1989). Participants were asked to speak aloud the final word in various sentences after reading the sentences. On average, they took less time to say the word in biased sentences like (1) than in sentences such as (2), indicating that "cocktails" had been activated, or primed, earlier in the sentence. But, unexpectedly, the times were as fast for sentences like (3) as they were for sentences like (1), even though the word "bartender" has no semantic connection to "cocktails" in (3).

(1) The boy watched the bartender serve the cocktails.
(2) The boy saw that the person liked the cocktails.
(3) The boy who watched the bartender served the cocktails.

Clearly, the semantic representation that yielded priming in (1) and (3) was not detailed enough to distinguish the difference in meaning between the two sentences. The representation was "good enough" to provide an interpretation that satisfied the comprehender, but not detailed enough to distinguish the important differences in who was doing what to whom.

RECENT STUDIES OF WHETHER INTERPRETATIONS ARE GOOD ENOUGH

In two series of studies, our lab has been investigating some situations in which good-enough, or noncompositional, processing may occur.

Misinterpretations of Garden-Path Sentences

One series (Christianson, Hollingworth, Halliwell, & Ferreira, 2001) addressed the straightforward question whether people delete from memory their initial misinterpretation of a sentence after reanalysis. When people were visually presented sentence (4), they initially took "the baby" to be the object of "dressed."

(4) While Anna dressed the baby played in the crib. (presented without commas)

As a result, readers spent a great deal of time processing the disambiguating word "played" and often reread the preceding material. Sentences such as this one are often termed *garden-path sentences*, because the first part of the sentence sends the language comprehension system in an ultimately wrong direction. The comprehender will have no difficulty with (4) if the clauses are separated by a comma or if the main clause is presented before the subordinate clause. In these cases, there is no temptation to take "the baby" to be the object of "dressed," and therefore the reader has no difficulty integrating "played."

It has generally been assumed that if comprehenders restructure their initial interpretation of (4) so as to make "the baby" the subject of the main clause, they will end up with an appropriate representation of the sentence's overall meaning. This assumption was tested by asking participants to respond to questions after reading (at their own pace) garden-path sentences or non-garden-path control versions of the same sentences (Christianson et al., 2001). The questions were of two sorts:

(5) Did the baby play in the crib?
(6) Did Anna dress the baby?

Question (5) assessed whether the phrase "the baby" was eventually taken to be the subject of "played." Recall that initially it is not; the syntactic processor makes "the baby" the object of "dressed," and so "played" ends up without a subject. Thus, successful syntactic restructuring requires that "the baby" be removed from that first clause and included in the second, making "yes" the correct answer to (5). Question (6) assessed whether comprehenders then adjusted the meaning of the sentence to correspond to that reanalysis: Under this reinterpretation, "the baby" is no longer the object of "dressed," and so the sentence means that Anna is dressing herself. Therefore, the participants should have said "no" in response to (6).

Participants were virtually 100% correct in responding that the baby played in the crib. Performance was equally good in the garden-path and non-garden-path conditions. Yet when the sentence led the comprehenders down a syntactic garden path, they were inaccurate in answering (6). That is, people initially took "the baby" to be the object of "dressed." Then, they restructured the sentence to make "the baby" the subject of "played," but they persisted in thinking that the baby was being dressed. People who read the non-garden-path control version, however, almost always correctly replied that Anna did not dress the baby. In summary, the initial misinterpretation lingered and caused comprehenders to end up with a representation in which "the baby" was both the subject of "played" and the object of "dressed." This is clear evidence that the meaning people obtain for a sentence is often not a reflection of its true content.

Misinterpretations of Passive Sentences

The other series of experiments (Ferreira & Stacey, 2000) was designed to investigate an even more basic question: Are people ever tricked by simple, but implausible, passive sentences? Consider an active sentence like (7). People have little trouble obtaining its implausible meaning. In contrast, the passive sentence (10) is much more difficult to understand, and one's impression is that it is hard to keep straight whether the dog is the perpetrator or the victim in the scenario.

(7) The man bit the dog.
(8) The man was bitten by the dog.
(9) The dog bit the man.
(10) The dog was bitten by the man.

In one experiment (Ferreira & Stacey, 2000), participants read sentences like (7) through (10) and were instructed to indicate whether the event described in each sentence was plausible. For the active sentences, people were almost always correct. However, they called passive sentences like (10) plausible more than 25% of the time. In another experiment, participants heard one of these four sentences and then identified either the agent or the patient of the action. Again, people were accurate with all sentences except (10). Thus, when people read or hear a passive sentence, they use their knowledge of the world to figure out who is doing what to whom. That interpretation reflects the content words of the sentence more than its compositional, syntactically derived meaning. It is as if people use a semantic heuristic rather than syntactic algorithms to get the meaning of difficult passives.

OUR GOOD-ENOUGH APPROACH

The linguistic system embodies a number of powerful mechanisms designed to enable the comprehender to obtain the meaning of a sentence that was intended by the speaker. The system uses mechanisms such as syntactic analysis to achieve this aim. Syntactic structure allows the comprehender to compute algorithmically who did what to whom, because it allows thematic roles such as agent to be bound to the individual words of the sentence. The challenges in comprehension, however, are twofold. First, as the earliest work in cognitive psychology revealed, the structure built by the language processor is fragile and decays rapidly (Sachs, 1967). The representation needs almost immediate support from context or from schemas (i.e., general frameworks used to organize details on the basis of previous experience). In other words, given (10), syntactic mechanisms deliver the proper interpretation that the dog is the patient and the man is the agent; but the problem is that the delicate syntactic structure needs reinforcement. Schemas in long-term memory cannot provide that support, and so the source of corroboration must be context. Quite likely, then, sentences like this would be correctly understood in normal conversation, because the overall communicative context would support the interpretation. The important concept is that the linguistic representation itself is not robust, so that if it is not reinforced, a merely good-enough interpretation may result.

The second challenge to the linguistic system is that it must cope with potentially interfering information. The garden-path studies show that an initial incorrect representation of a sentence lingers and interferes with obtaining the correct meaning for the sentence. In the case of implausible passive sentences, information from schemas in long-term memory causes interference. As a result, people end up believing that (10) means what their schema tells them rather than what the output of the syntactic algorithms mandates. This interfering information must be inhibited for comprehension to be successful.

FUTURE DIRECTIONS

Experiments are under way to examine the characteristics of the memory representations for garden-path sentences, and to focus on how misinformation is suppressed during successful comprehension. The studies on passives are intriguing because they demonstrate that complex syntactic structures can be misinterpreted, but what makes a structure likely to be misinterpreted? One of the experiments (Ferreira & Stacey, 2000) demonstrated that the surface frequency of the sentence form is not critical to determining difficulty. People were as accurate with sentences such as "It was the man who bit the dog" as they were with common active sentences, even though the former structure is rare. One possible explanation for why the passive structure is difficult to comprehend is that passives require semantic roles to be assigned in an atypical order: patient before agent. This hypothesis can be addressed by examining languages that permit freer word order than does English. We are currently focusing on the aboriginal Native American language Odawa, which orthogonally crosses voice and word order—that is, an active sentence may have the patient either before or after the agent, as may a passive sentence. Thus, Odawa provides a unique opportunity for us to study the factors that cause linguistic representations to be particularly fragile and vulnerable to influence from schemas.

The good-enough approach also leads us in several other less traditional directions. For example, speech disfluencies that occur during conversation include pauses filled with "uh" or "um," repeated words, repairs that modify or replace earlier material, and false starts (utterance fragments that are begun and abandoned). Disfluencies will often yield a string of words that violates grammatical principles. Nevertheless, comprehenders seem able to process such strings efficiently, and it is not clear how interpretation processes are affected by these disfluencies. Are abandoned fragments incorporated into the semantic representation of a sentence? Our work on misinterpretations of garden-path sentences suggests that the answer could well be yes. In the same way that the incorrect interpretation of a garden-path sentence lingers even though its underlying structure is ultimately corrected, an interpretation built upon an ultimately abandoned fragment (e.g., "Turn left—I mean right at the stop sign") might persist in the comprehender's overall representation.

We are also investigating whether syntactically ambiguous sentences such as (11) and (12) are given incomplete syntactic representations. A recent study found that people were faster at reading sentences like (11), for which the attachment of the relative clause is semantically ambiguous, than at reading semantically unambiguous versions like (12) (Traxler, Pickering, & Clifton, 1998).

(11) The son of the driver that had the mustache was pretty cool.
(12) The car of the driver that had the mustache was pretty cool.

One proposed explanation for this finding is that the syntactic representation in the ambiguous case remains underspecified. That is, perhaps the language processor does not bother to attach the relative clause "that had the mustache" to either "son" or "driver" because it does not have enough information to support one interpretation over the other.

More generally, the good-enough approach to language comprehension invites a more naturalistic perspective on how people understand utterances than has been adopted in psycholinguistics up to this point. Psycholinguists have focused on people's ability to understand individual sentences (or short texts) in almost ideal circumstances. In laboratories, stimuli are (usually) shown visually in quiet rooms that offer no distractions. The results that have emerged from this work are central to any theory of comprehension, but examination of only those conditions will not yield a complete story. Outside the laboratory, utterances are often difficult to hear because of background noise; dialect and idiolect differences as well as competing sounds can make it difficult for the hearer to extract every word from an utterance; and speakers often produce utterances with disfluencies and outright errors that the processing system must handle somehow. We have shown in our research that, even in the ideal conditions of the laboratory, comprehension is more shallow and incomplete than psycholinguists might have suspected. In the real world, interpretations are even more likely to be "just good enough."

Perhaps good-enough interpretations help the language system coordinate listening and speaking during conversation. Usually when people talk to one another, turns are not separated by gaps. Therefore, comprehension and production processes must operate simultaneously. The goal of the comprehension system might be to deliver an interpretation that is just good enough to allow the production system to generate an appropriate response; after all, it is the response that is overt and that determines the success of the participants' joint activity. An adequate theory of how language is understood, then, will ultimately have to take into account the dynamic demands of real-time conversation.

Recommended Reading

Christianson, K., Hollingworth, A., Halliwell, J., & Ferreira, F. (2001). (See References)
Clark, H.H. (1996). *Using language*. Cambridge, England: Cambridge University Press.
Clifton, C., Jr. (2000). Evaluating models of human sentence processing. In M.W. Crocker, M. Pickering, & C. Clifton, Jr. (Eds.), *Architectures and mechanisms for language processing* (pp. 31–55). Cambridge, England: Cambridge University Press.
Tanenhaus, M.K., Spivey-Knowlton, M.J., Eberhard, K.M., & Sedivy, J.E. (1995). Integration of visual and linguistic information in spoken language comprehension. *Science, 268*, 632–634.

Note

1. Address correspondence to Fernanda Ferreira, 129 Psychology Research Building, Michigan State University, East Lansing, MI 48824-1117; e-mail: fernanda@eyelab.msu.edu.

References

Barton, S.B., & Sanford, A.J. (1993). A case study of anomaly detection: Shallow semantic processing and cohesion establishment. *Memory & Cognition, 21*, 477–487.
Christianson, K., Hollingworth, A., Halliwell, J., & Ferreira, F. (2001). Thematic roles assigned along the garden path linger. *Cognitive Psychology, 42*, 368–407.
Duffy, S.A., Henderson, J.M., & Morris, R.K. (1989). Semantic facilitation of lexical access during sentence processing. *Journal of Experimental Psychology: Learning, Memory, and Cognition, 15*, 791–801.
Erickson, T.A., & Mattson, M.E. (1981). From words to meaning: A semantic illusion. *Journal of Verbal Learning and Verbal Behavior, 20*, 540–552.
Ferreira, F., & Clifton, C., Jr. (1986). The independence of syntactic processing. *Journal of Memory and Language, 25*, 348–368.
Ferreira, F., & Stacey, J. (2000). *The misinterpretation of passive sentences*. Manuscript submitted for publication.
Fillenbaum, S. (1974). Pragmatic normalization: Further results for some conjunctive and disjunctive sentences. *Journal of Experimental Psychology, 102*, 574–578.
Frazier, L. (1978). *On comprehending sentences: Syntactic parsing strategies*. Unpublished doctoral dissertation, University of Connecticut, Storrs.
MacDonald, M.C., Pearlmutter, N.J., & Seidenberg, M.S. (1994). The lexical nature of syntactic ambiguity resolution. *Psychological Review, 101*, 676–703.
Sachs, J.S. (1967). Recognition memory for syntactic and semantic aspects of connected discourse. *Perception & Psychophysics, 2*, 437–442.
Traxler, M.J., Pickering, M.J., & Clifton, C., Jr. (1998). Adjunct attachment is not a form of ambiguity resolution. *Journal of Memory and Language, 39*, 558–592.
Trueswell, J., Tanenhaus, M., & Garnsey, S. (1994). Semantic influences on parsing: Use of thematic role information in syntactic disambiguation. *Journal of Memory and Language, 33*, 285–318.

Critical Thinking Questions

1. A consequence of the "good enough" approach is that errors of comprehension *will* occur. If language comprehension is "good enough" these errors should rarely cause significant problems. Think carefully about how often you

misunderstand someone or someone misunderstands you; do you believe that such harmless misunderstandings often occur in your day-to-day life?

2. Do you think that there are other cognitive processes that have this "good enough" quality? For example, is your visual system perfect? Is your motor system perfect? If not, what sorts of errors do they make, and in what sense are these systems (or others) "good enough?"

3. It is surprising that we fail to fully update our knowledge even when we *know* we have made a comprehension error, as in the sample sentence of the baby in the crib being the one getting dressed. Can you think of any reason that one would fail to correct this misinformation? Does this failure mean that incorrect representations are still good enough? If so, what would *not* be good enough?

Situation Models: The Mental Leap Into Imagined Worlds

Rolf A. Zwaan[1]
Department of Psychology, Florida State University, Tallahassee, Florida

Abstract

Situation models are mental representations of the state of affairs described in a text rather than of the text itself. Much of the research on situation models in narrative comprehension suggests that comprehenders behave as though they are in the narrated situation rather than outside of it. This article reviews some of this evidence and provides an outlook on future developments.

Keywords

situation models; language; comprehension

> When reading a fictional text, most readers feel they are in the middle of the story, and they eagerly or hesitantly wait to see what will happen next. Readers get inside of stories and vicariously experience them. They feel happy when good things occur, worry when characters are in danger, feel sad, and may even cry when misfortune strikes. While in the middle of a story, they are likely to use past tense verbs for events that have already occurred, and future tense for those that have not. (Segal, 1995, p. 65)

In the 1980s, researchers proposed that understanding a story, or any text for that matter, involves more than merely constructing a mental representation of the text itself. Comprehension is first and foremost the construction of a mental representation of what that text is about: a situation model. Thus, situation models are mental representations of the people, objects, locations, events, and actions described in a text, not of the words, phrases, clauses, sentences, and paragraphs of a text. The situation-model view predicts that comprehenders are influenced by the nature of the situation that is described in a text, rather than merely by the structure of the text itself.

As a first illustration, consider the following sentences: *Mary baked cookies but no cake* versus *Mary baked cookies and cake*. Both sentences mention the word *cake* explicitly, but only the second sentence refers to a situation in which a cake is actually present. If comprehenders construct situation models, the concept of cake should be more available to them when the cake is in the narrated situation than when it is not, despite the fact that the word *cake* appears in both sentences. Consistent with this prediction, students who read (from a computer screen) short narratives containing sentences such as these recognized words (presented immediately after each text) more quickly when the denoted object was actually present in the narrated situation than when it was not (MacDonald & Just, 1989).

G.A. Radvansky and I have recently reviewed the extensive literature on situation models (Zwaan & Radvansky, 1998). Here, I focus specifically on the evidence pertaining to situation models as vicarious experiences in narrative comprehension. When we place ourselves in a situation, we have a certain spatial, temporal, and psychological "vantage" point from which we vicariously experience the events. Such a perspective has been termed a *deictic center*, and the shift to this perspective a *deictic shift* (Duchan, Bruder, & Hewitt, 1995). In everyday life, we are typically aware of our location and time. We are also aware of our current goals. We are aware of people in our environment and their goals and emotions. And we are aware of objects that are relevant to our goals. This is a useful first approximation of what should be relevant to a deictic center.

SPACE

People exist in, move about in, and interact with environments. Situation models should represent relevant aspects of these environments. Very often (but not necessarily), objects that are spatially close to us are more relevant than more distant objects. Therefore, one would expect the same for situation models. Consistent with this idea, comprehenders are slower to recognize words denoting objects distant from a protagonist than those denoting objects close to the protagonist (Glenberg, Meyer, & Lindem, 1987).

When comprehenders have extensive knowledge of the spatial layout of the setting of the story (e.g., a building), they update their representations according to the location and goals of the protagonist. They have the fastest mental access to the room that the protagonist is currently in or is heading to. For example, they can more readily say whether or not two objects are in the same room if the room mentioned is one of these rooms than if it is some other room in the building (e.g., Morrow, Greenspan, & Bower, 1987). This makes perfect sense intuitively; these are the rooms that would be relevant to us if we were in the situation.

People's interpretation of the meaning of a verb denoting movement of people or objects in space, such as *to approach*, depends on their situation models. For example, comprehenders interpret the meaning of *approach* differently in *The tractor is just approaching the fence* than in *The mouse is just approaching the fence*. Specifically, they interpret the distance between the figure and the landmark as being longer when the figure is large (tractor) compared with when it is small (mouse). The comprehenders' interpretation also depends on the size of the landmark and the speed of the figure (Morrow & Clark, 1988). Apparently, comprehenders behave as if they are actually standing in the situation, looking at the tractor or mouse approaching a fence.

TIME

We assume by default that events are narrated in their chronological order, with nothing left out. Presumably this assumption exists because this is how we experience events in everyday life. Events occur to us in a continuous flow, sometimes in close succession, sometimes in parallel, and often partially overlapping. Language allows us to deviate from chronological order, however. For example, we can say, "Before the psychologist submitted the manuscript, the journal changed its policy." The psychologist submitting the manuscript is reported first, even though it was the last of the two events to occur. If people construct a situation model, this sentence should be more difficult to process than its chronological counterpart (the same sentence, but beginning with "After"). Recent neuroscientific evidence supports this prediction. Event-related brain potential (ERP) measurements[2] indicate that "before" sentences elicit, within 300 ms, greater negativity than "after" sentences. This difference in potential is primarily located in the left-anterior part of the brain and is indicative of greater cognitive effort (Münte, Schiltz, & Kutas, 1998).

In real life, events follow each other seamlessly. However, narratives can have temporal discontinuities, when writers omit events not relevant to the plot. Such temporal gaps, typically signaled by phrases such as *a few days later*, are quite common in narratives. Nonetheless, they present a departure from everyday experience. Therefore, time shifts should lead to (minor) disruptions of the comprehension process. And they do. Reading times for sentences that introduce a time shift tend to be longer than those for sentences that do not (Zwaan, 1996).

All other things being equal, events that happened just recently are more accessible to us than events that happened a while ago. Thus, in a situation model, *enter* should be less accessible after *An hour ago, John entered the building* than after *A moment ago, John entered the building*. Recent probe-word recognition experiments support this prediction (e.g., Zwaan, 1996).

GOALS AND CAUSATION

If we have a goal that is currently unsatisfied, it will be more prominent in our minds than a goal that has already been accomplished. For example, my goal to assist my wife in preparing for a party at our house tonight is currently more active in my mind than my goal to write a review of a manuscript if I finished the review this morning. Once a goal has been accomplished, there is no need for me to keep it on my mental desktop. Thus, if a protagonist has a goal that has not yet been accomplished, that goal should be more accessible to the comprehender than a goal that was just accomplished by the protagonist. In line with this prediction, goals yet to be accomplished by the protagonist were recognized more quickly than goals that were just accomplished (Trabasso & Suh, 1993).

We are often able to predict people's future actions by inferring their goals. For example, when we see a man walking over to a chair, we assume that he wants to sit, especially when he has been standing for a long time. Thus, we might generate the inference "He is going to sit." Keefe and McDaniel (1993) presented subjects with sentences like *After standing through the 3-hr debate, the tired speaker walked over to his chair (and sat down)* and then with probe words (e.g., *sat*, in this case). Subjects took about the same amount of time to name *sat* when the clause about the speaker sitting down was omitted and when it was included. Moreover, naming times were significantly faster in both of these conditions than in a control condition in which it was implied that the speaker remained standing.

As we interact with the environment, we have a strong tendency to interpret event sequences as causal sequences. It is important to note that, just as we infer goals, we have to infer causality; we cannot perceive it directly. Singer and his colleagues (e.g., Singer, Halldorson, Lear, & Andrusiak, 1992) have investigated how readers use their world knowledge to validate causal connections between narrated events. Subjects read sentence pairs, such as 1a and then 1b or 1a′ and then 1b, and were subsequently presented with a question like 1c:

(1a) Mark poured the bucket of water on the bonfire.

(1a′) Mark placed the bucket of water by the bonfire.

(1b) The bonfire went out.

(1c) Does water extinguish fire?

Subjects were faster in responding to 1c after the sequence 1a–1b than after 1a′–1b. According to Singer, the reason for this is that the knowledge that water extinguishes fire was activated to validate the events described in 1a–1b. However, because this knowledge cannot be used to validate 1a′–1b, it was not activated when subjects read that sentence pair.

PEOPLE AND OBJECTS

Comprehenders are quick to make inferences about protagonists, presumably in an attempt to construct a more complete situation model. Consider, for example, what happens after subjects read the sentence *The electrician examined the light fitting*. If the following sentence is *She took out her screwdriver*, their reading speed is slowed down compared with when the second sentence is *He took out his screwdriver*. This happens because *she* provides a mismatch with the stereotypical gender of an electrician, which the subjects apparently infer while reading the first sentence (Carreiras, Garnham, Oakhill, & Cain, 1996).

Comprehenders also make inferences about the emotional states of characters. For example, if we read a story about Paul, who wants his brother Luke to be good in baseball, the concept of "pride" becomes activated in our mind when we read that Luke receives the Most Valuable Player Award (Gernsbacher, Goldsmith, & Robertson, 1992). Thus, just as in real life, we make inferences about people's emotions when we comprehend stories.

Just as we empathize with real people, we seem to empathize with story protagonists. Comprehenders' preferences for a particular outcome of a story interfere with the verification of previously known information about the actual outcome of the story. For example, comprehenders had difficulty verifying that "Margaret made her flight" when they had learned previously that Margaret's plane would plunge into the sea shortly after takeoff (Allbritton & Gerrig, 1991). Allbritton and Gerrig hypothesized that during reading, comprehenders generated *participatory* responses (e.g., "I hope she will miss the flight") that interfered with their verification performance.

THE FUTURE OF SITUATION MODELS

How close are we to a scientific account of the vicarious experiences described in the epigraph to this article? Advances are to be expected on two fronts. On the theoretical front, there will be discussion of the proper representational format for situation models. Researchers, most notably Kintsch (1998), have proposed computer models of how people construct situation models. The question has been raised recently as to whether such computer-based models can account for the full complexity of situation-model construction (and human cognition in general), or whether a biologically oriented approach has more explanatory power (e.g., Barsalou, in press). On the methodological front, the repertoire of cognitive tasks is being supplemented with measures of brain activity. Initial findings provide converging evidence (e.g., Münte et al., 1998).

To summarize, many aspects of narrated situations have already been shown to affect our understanding of stories. However, there is still a great deal that must be learned before we have a good understanding of people's fascinating ability to make a mental leap from their actual situation, reading a book on the couch, to an often fictional situation at a different time and place. Recent theoretical and methodological developments give reason to be optimistic about this endeavor.

Recommended Reading

Duchan, J.F., Bruder, G.A., & Hewitt, L.E. (Eds.). (1995). (See References)

Graesser, A.C., Millis, K.K., & Zwaan, R.A. (1997). Discourse comprehension. *Annual Review of Psychology*, 48, 163–189.

Johnson-Laird, P.N. (1983). *Mental models: Towards a cognitive science of language, inference, and consciousness*. Cambridge, MA: Harvard University Press.

Kintsch, W. (1998). (See References)

Zwaan, R.A., & Radvansky, G.A. (1998). (See References)

Notes

1. Address correspondence to Rolf A. Zwaan, Department of Psychology, Florida State University, Tallahassee, FL 32306-1270; e-mail: zwaan@psy.fsu.edu; World Wide Web: http://freud.psy.fsu.edu:80/~zwaan/.

2. ERPs are modulations of electrical activity in the brain that occur as a result of the processing of external stimuli.

References

Allbritton, D.W., & Gerrig, R.J. (1991). Participatory responses in prose understanding. *Journal of Memory and Language*, 30, 603–626.

Barsalou, L.W. (in press). Perceptual symbol systems. *Behavioral and Brain Sciences*.

Carreiras, M., Garnham, A., Oakhill, J., & Cain, K. (1996). The use of stereotypical gender information in constructing a mental model: Evidence from English and Spanish. *Quarterly Journal of Experimental Psychology*, 49A, 639–663.

Duchan, J.F., Bruder, G.A., & Hewitt, L.E. (Eds.). (1995). *Deixis in narrative: A cognitive science perspective*. Hillsdale, NJ: Erlbaum.

Gernsbacher, M.A., Goldsmith, H.H., & Robertson, R.W. (1992). Do readers mentally represent characters' emotional states? *Cognition and Emotion*, 6, 89–111.

Glenberg, A.M., Meyer, M., & Lindem, K. (1987). Mental models contribute to foregrounding during text comprehension. *Journal of Memory and Language*, 26, 69–83.

Keefe, D.E., & McDaniel, M.A. (1993). The time course and durability of predictive inferences. *Journal of Memory and Language*, 32, 446–463.

Kintsch, W. (1998). *Comprehension: A paradigm for cognition*. Cambridge, England: Cambridge University Press.

MacDonald, M.C., & Just, M.A. (1989). Changes in activation level with negation. *Journal of Experimental Psychology: Learning, Memory, and Cognition, 15*, 633–642.

Morrow, D.G., & Clark, H.H. (1988). Interpreting words in spatial descriptions. *Language and Cognitive Processes, 3*, 275–291.

Morrow, D.G., Greenspan, S.L., & Bower, G.H. (1987). Accessibility and situation models in narrative comprehension. *Journal of Memory and Language, 26*, 165–187.

Münte, T.F., Schiltz, K., & Kutas, M. (1998). When temporal terms belie conceptual order. *Nature, 395*, 71–73.

Segal, E.M. (1995). Cognitive-phenomenological theory of fictional narrative. In J.F. Duchan, G.A. Bruder, & L.E. Hewitt (Eds.), *Deixis in narrative: A cognitive science perspective* (pp. 61–78). Hillsdale, NJ: Erlbaum.

Singer, M., Halldorson, M., Lear, J.C., & Andrusiak, P. (1992). Validation of causal bridging inferences. *Journal of Memory and Language, 31*, 507–524.

Trabasso, T., & Suh, S. (1993). Understanding text: Achieving explanatory coherence through online inferences and mental operations in working memory. *Discourse Processes, 16*, 3–34.

Zwaan, R.A. (1996). Processing narrative time shifts. *Journal of Experimental Psychology: Learning, Memory, and Cognition, 22*, 1196–1207.

Zwaan, R.A., & Radvansky, G.A. (1998). Situation models in language comprehension and memory. *Psychological Bulletin, 123*, 162–185.

Critical Thinking Questions

1. Zwaan discusses situation models for stories. Do you think that we create situation models when we read other sorts of texts, for example, a scientific article, or a historical account of a civil war battle? How about the instructions for constructing a piece of furniture?

2. Some researchers have suggested that the process in our minds that creates situation models when we read actually evolved for dealing with social situations. In other words, we create situation models to keep track of what others are doing in real life, not just when we read, and we often adopt others' point of view when we do so. From your own experience, evaluate this idea, paying special attention to those properties of a situation model that Zwaan says are important.

3. Does our knowledge of situation models help us to predict what people will say is a well-formed story? Does it help us to predict what sort of story people will find surprising, exciting, or interesting?

Minds and Brains

Cognitive psychology is the study of the individual mind: What is its architecture and how does it function? Addressing these questions inevitably leads to closely related questions. What is the relationship of the mind and brain? How do animal minds differ from human minds? Considering such questions may help us to better understand the human mind, or even to reveal problems in our science that had previously gone undetected.

Perhaps the most obvious (and most timely) of these problems is the relationship of mind and brain. Enormous strides have been made in neuroscience in the last twenty years. How may such data be integrated into a study of the mind? More important, do they render study of the mind irrelevant? Perhaps cognitive psychologists are merely playing tic-tac-toe until the neuroscientists get it right. According to Miller and Keller in "Psychology and Neuroscience: Making Peace," this view—that psychology can be reduced to neuroscience—is flawed, although it is a common perception even among psychologists.

Neuroscience will not someday supplant cognitive psychology—but will it help us now? The answer is a clear "yes," as exemplified by Roser and Gazzaniga's article titled "Automatic Brains—Interpretative Minds." The problem they address is this: most cognitive psychologists hold a modular view of the mind. That means they believe that the mind is composed of different processing modules, each performing a different job, and each relatively autonomous of the others (although they do communicate). For example, different processing modules within the visual system appear to process color information, shape, motion, and so on. This view of the mind is supported by data from neuroscience showing different anatomic substrates for these different cognitive modules. Modularity, however, gives rise to a question: If the mind is composed of separate processing modules, why is conscious experience unitary, rather than fragmented? Although the mind appears to tear the world apart into different components that are processed separately, the products of these processes are somehow knit together again. Roser and Gazzaniga argue that a process in the left hemisphere constructs and interprets a personal narrative from these disparate data, affording a single, coherent experience.

In trying to understand the human mind, cognitive psychologists turn not only to the human brain, but also to animal minds. Scientists often begin with simple systems and, once those systems are understood, move onto more complex systems. In studying non-human primates, we might suspect that we are observing simpler versions of ourselves. In "The Mentality of Apes Revisited," Povinelli and Bering argue that this attitude is pervasive in psychology, and that it is dead wrong. There is not an "evolutionary ladder" with humans at the pinnacle and other animals at various rungs below us. Evolution does not create a ladder with a single dimension of human-like intelligence, with different animals endowed with more or less. The product of evolution is diversity of mental function, with each animal possessing different cognitive abilities, suited to its environment. Povinelli and Bering believe that most psychologists have made a fundamental mistake when thinking about the cognition of other species.

Most humans have only a passing interest in what apes know, but we are all forced to make judgments about what other humans know. For example, suppose that I want to tell you about an interesting news story I heard on the radio about current events in Central America. I need to judge how much you likely know about the topic already. If I underestimate, I will provide background that you already know and I will likely bore you. If I overestimate your background knowledge, the story will likely be incomprehensible to you. How do I judge what you know? In "The Projective Way of Knowing: A Useful Heuristic that Sometimes Misleads," Nickerson reviews evidence indicating that we use a very simple rule of thumb in these situations: in the absence of other information we assume that others know what we know. In other words we assume that other brains, or other minds, are like our own.

Psychology and Neuroscience: Making Peace

Gregory A. Miller[1] and Jennifer Keller
Department of Psychology, University of Illinois at Urbana-Champaign, Champaign, Illinois

Abstract

There has been no historically stable consensus about the relationship between psychological and biological concepts and data. A naively reductionist view of this relationship is prevalent in psychology, medicine, and basic and clinical neuroscience. This view undermines the ability of psychology and related sciences to achieve their individual and combined potential. A nondualistic, nonreductionist, noninteractive perspective is recommended, with psychological and biological concepts both having central, distinct roles.

Keywords

psychology; biology; neuroscience; psychopathology

With the Decade of the Brain just ended, it is useful to consider the impact that it has had on psychological research and what should come next. Impressive progress occurred on many fronts, including methodologies used to understand the brain events associated with psychological functions. However, much controversy remains about where biological phenomena fit into psychological science and vice versa. This controversy is especially pronounced in research on psychopathology, a field in which ambitious claims on behalf of narrowly conceived psychological or biological factors often arise, but this fundamental issue applies to the full range of psychological research. Unfortunately, the Decade of the Brain has fostered a naively reductionist view that sets biology and psychology at odds and often casts psychological events as unimportant epiphenomena. We and other researchers have been developing a proposal that rejects this view and provides a different perspective on the relationship between biology and psychology.

A FAILURE OF REDUCTIONISM

A term defined in one domain is characterized as *reduced* to terms in another domain (called the reduction science) when all meaning in the former is captured in the latter. The reduced term thus becomes unnecessary. If, for example, the meaning of the (traditionally psychological) term "fear" is entirely representable in language about a brain region called the amygdala, one does not need the (psychological) term "fear," or one can redefine "fear" to refer merely to a particular biological phenomenon.

Impressive progress in the characterization of neural circuits typically active in (psychologically defined) fear does not justify dismissing the concept or altering the meaning of the term. The phenomena that "fear" typically refers to include a functional state (a way of being or being prepared to act), a cognitive processing bias, and a variety of judgments and associations all of which are conceived psychologically (Miller & Kozak, 1993). Because "fear" means more than a given type of neural activity, the concept of fear is not reducible to neural activity. Researchers are learning a great deal about the biology of fear—and the psychology of fear—from studies of the amygdala (e.g., Lang, Davis, & Öhman, in press), but this does not mean that fear is activity in the amygdala. That is simply not the meaning of the term. "Fear" is not reducible to biology.

This logical fact is widely misunderstood, as evidenced in phrases such as "underlying brain dysfunction" or "neurochemical basis of psychopathology." Most remarkably, major portions of the federal research establishment have recently adopted a distinctly nonmental notion of mental health, referring to "the biobehavioral factors which may underly [sic] mood states" (National Institute of Mental Health, 1999). Similarly, a plan to reorganize grant review committees reflects "the context of the biological question that is being investigated" (National Institutes of Health, 1999, p. 2). Mental health researchers motivated by psychological or sociological questions apparently should take their applications elsewhere.

More subtly problematic than such naive reductionism are terms, such as "biobehavioral marker" or "neurocognitive measure," that appear to cross the boundary between psychological and biological domains. It is not at all apparent what meaning the "bio" or "neuro" prefix adds in these terms, as typically the data referred to are behavioral. Under the political pressures of the Decade of the Brain, psychologists were tempted to repackage their phenomena to sound biological, but the relationship of psychology and biology cannot be addressed by confusing them.

WHOSE WORK IS MORE FUNDAMENTAL?

Such phrases often appear in contexts that assume that biological phenomena are somehow more fundamental than psychological phenomena. Statements that psychological events are nothing more than brain events are clearly logical errors (see the extensive analysis by Marr, 1982). More cautious statements, such as that psychological events "reflect" or "arise from" brain events, are at best incomplete in what they convey about the relationship between psychology and biology. It is not a property of biological data that they "underlie" psychological data. A given theory may explicitly propose such a relationship, but it must be treated as a proposal, not as a fact about the data. Biological data provide valuable information that may not be obtainable with self-report or overt behavioral measures, but biological information is not inherently more fundamental, more accurate, more representative, or even more objective.

The converse problem also arises—psychology allegedly "underlying" or being more fundamental than biology. There is a long tradition of ignoring biological phenomena in clinical psychology. As Zuckerman (1999) noted, "One thing that both behavioral and post-Freudian psychoanalytic theories had in common was the conviction that learning and life experiences alone could account for all disorders" (p. 413). In those traditions, it is psychology that "underlies" biology, not the converse. Biology is seen as merely the implementation of psychology, and psychology is where the intellectually interesting action is. Cognitive theory can thus evolve without the discipline of biological plausibility. As suggested at the midpoint of the Decade of the Brain (Miller, 1995), such a view would justify a Decade of Cognition.

Such a one-sided emphasis would once again be misguided. Anderson and Scott (1999) expressed concern that "the majority of research in the health sciences occurs within a single level of analysis, closely tied to specific disciplines" (p. 5), with most psychologists studying phenomena only in terms of behavior. We advocate not that every study employ both psychological and biological methods, but that researchers not ignore or dismiss relevant literatures, particularly in the conceptualization of their research.

Psychological and biological approaches offer distinct types of data of potentially equal relevance for understanding psychological phenomena. For example, we use magnetoencephalography (MEG) recordings of the magnetic fields generated by neural activity to identify multiple areas of brain tissue that are generating what is typically measured electrically at the scalp (via electroencephalography, or EEG) as the response of the brain associated with cognitive tasks (Cañive, Edgar, Miller, & Weisend, 1999). One of the most firmly established biological findings in schizophrenia is a smaller than normal brain response called the P300 component (Ford, 1999), and there is considerable consensus on the functional significance of P300 in the psychological domain. There is, however, no consensus on what neural generators produce the electrical activity or on what distinct functions those generators serve. Neural sources are often difficult to identify with confidence from EEG alone, whereas for biophysical reasons MEG (which shows brain function) coupled with structural magnetic resonance data (which show brain anatomy) promises localization as good as any other available noninvasive method. If researchers understand the distinct functional significance of various neural generators of P300, and if only some generators are compromised in schizophrenia, this will be informative about the nature of cognitive deficits in schizophrenia. Conversely, what researchers know about cognitive deficits will be informative about the function of the different generators.

MEG and EEG do not "underlie" and are not the "basis" of (the psychological phenomena that define) the functions or mental operations invoked in tasks associated with the P300 response. Neural generators implement functions, but functions do not have locations (Fodor, 1968). For example, a working memory deficit in schizophrenia could not be located in a specific brain region. The psychological and the neuromagnetic are not simply different "levels" of analysis, except in a very loose (and unhelpful) metaphorical sense. Neither underlies the other, neither is more fundamental, and neither explains away the other. There are simply two domains of data, and each can help to explicate the other because of the relationships theories propose.

Psychophysiological research provides many other examples in which the notion of "underlying" is unhelpful. Rather than attributing mood changes to activity in specific brain regions, why not attribute changes in brain activity to changes in mood? In light of EEG (Deldin, Keller, Gergen, & Miller, 2000) or behavioral (Keller et al., 2000) data on regional brain activity in depression, are people depressed because of low activity in left frontal areas of the brain, or do they have low activity in these areas because they are depressed? Under the present view, such a question, trying to establish causal relations *between* psychology and biology, is misguided. These are not empirical issues but logical and theoretical issues. They turn on the kind of relationship that psychological and biological concepts are proposed to have.

CLINICAL IMPLICATIONS

In psychopathology, one of the most unfortunate consequences of the naive competition between psychology and biology is the assumption that dysfunctions conceptualized biologically require biological interventions and that those conceptualized psychologically require psychological interventions. The best way to alter one system may be a direct intervention in another system. Even, for example, if the chemistry of catecholamines (chemicals used for communication to nerve, muscle, and other cells) were the best place to intervene in schizophrenia, it does not follow that a direct biological intervention in that system would be optimal. A variety of experiences that people construe as psychosocial prompt their adrenal glands to flood them with catecholamines. There are psychological interventions associated with this chemistry that can work more effectively or with fewer side effects than medications aimed directly at the chemistry.

Unfortunately, the assumption that disorders construed biologically warrant exclusively biological interventions influences not only theories of psychopathology but also available treatments. For example, major depression is increasingly viewed as a "chemical imbalance." If such (psychological) disorders are assumed to "be" biological, then medical insurers are more likely to fund only biological treatments. Yet Thase et al. (1997) found that medication and psychotherapy were equally effective in treating moderately depressed patients and that the combination of these treatments was more effective than either alone in treating more severely depressed patients. Hollon (1995) discussed how negative life events may alter biological factors that increase risk for depression. Meany (1998) explained how the psychological environment can affect gene activity. The indefensible conceptualization of depression solely as a biological disorder prompts inappropriately narrow (biological) interventions. Thus, treatment as well as theory is hampered by naive reductionism.

WHAT TO DO?

"Underlying" (implying one is more fundamental than the other) is not a satisfactory way to characterize the relationship between biological and psychological concepts. We recommend characterizing the biological as "implementing" the psychology—that is, we see cognition and emotion as implemented in neural systems. Fodor (1968) distinguished between *contingent* and *necessary* identity in the relationship between psychological and biological phenomena. A person in any given psychological state is momentarily in some biological state as well: There is a *contingent* identity between the psychological and the biological at that moment. The psychological phenomenon implemented in a given neural circuit is not the same as, is not accounted for by, and is not reducible to that circuit.

There is an indefinite set of potential neural implementations of a given psychological phenomenon. Conversely, a given neural circuit might implement different psychological functions at different times or in different individuals. Thus, there is no *necessary* identity between psychological states and brain states. Distinct psychological and biological theories are needed to explain their respective domains, and additional theoretical work is needed to relate them.

Nor is it viable (though it is common) to say that psychological and biological phenomena "interact." Such a claim begs the question of how they interact and even what it means to interact. The concept of the experience of "red" does not "interact" with the concept of photon-driven chemical changes in the retina and their neural sequelae. One may propose that those neural sequelae implement the perceptual experience of "red," but "red" *means* not the neural sequelae, but something psychological—a perception.

Biology and psychology often are set up as competitors for public mind-share, research funding, and scientific legitimacy. We are not arguing for a psychological explanation of cognition and emotion *instead of* a biological explanation. Rather, we are arguing against framing biology and psychology in a way that forces a choice between those kinds of explanations. The hyperbiological bias ascendant at the end of the 20th century was no wiser and no more fruitful than the hyperpsychological bias of the behaviorist movement earlier in the 20th century. Scientists can avoid turf battles by approaching the relationship between the psychological and the biological as fundamentally theoretical, not empirical. Working out the biology will not make psychology obsolete, any more than behaviorism rendered biology obsolete. Scientists can avoid dualism by avoiding interactionism (having two distinct domains in a position to interact implies separate realities, hence dualism). Psychological and biological domains can be viewed as logically distinct but not physically distinct, and hence neither dualistic nor interacting. Psychological and biological concepts are not merely different terms for the same phenomena (and thus not reducible in either direction), and psychological and biological explanations are not explanations of the same things. If one views brain tissue as implementing psychological functions, the expertise of cognitive science is needed to characterize those functions, and the expertise of neuroscience is needed to study their implementation. Each of those disciplines will benefit greatly from the other, but neither encompasses, reduces, or underlies the other.

Fundamentally psychological concepts require fundamentally psychological explanations. Stories about biological phenomena can richly inform, but not supplant, those explanations. Yet when psychological events unfold, they are implemented in biology, and those implementations are extremely important to study as well. For example, rather than merely pursuing, in quite separate literatures, anomalies in either expressed emotion or biochemistry, research on schizophrenia should investigate biological mechanisms involved in expressed-emotion phenomena. Similarly, the largely separate literatures on biological and psychosocial mechanisms in emotion should give way to conceptual and methodological collaboration. Research in the next few decades will need not only the improving spatial resolution of newer brain-imaging technologies and the high temporal resolution of established brain-imaging technologies, but also the advancing cognitive resolution of the best psychological science.

Recommended Reading

Anderson, N.B., & Scott, P.A. (1999). (See References)

Cacioppo, J.T., & Berntson, G.G. (1992). Social psychological contributions to the Decade of the Brain. *American Psychologist, 47,* 1019–1028.

Kosslyn, S.M., & Koenig, O. (1992). *Wet mind: The new cognitive neuroscience.* New York: Free Press.

Miller, G.A. (1996). Presidential address: How we think about cognition, emotion, and biology in psychopathology. *Psychophysiology, 33,* 615–628.

Ross, C.A., & Pam, A. (1995). *Pseudoscience in biological psychiatry: Blaming the body.* New York: Wiley.

Acknowledgments—The authors' work has been supported in part by National Institute of Mental Health Grants R01 MH39628, F31 MH11758, and T32 MH19554; by the Department of Psychiatry of Provena Covenant Medical Center; and by the Research Board, the Beckman Institute, and the Departments of Psychology and Psychiatry of the University of Illinois at Urbana-Champaign. The authors appreciate the comments of Howard Berenbaum, Patricia Deldin, Wendy Heller, Karen Rudolph, Judith Ford, Michael Kozak, Sumie Okazaki, and Robert Simons on an earlier draft.

Note

1. Address correspondence to Gregory A. Miller, Departments of Psychology and Psychiatry, University of Illinois, 603 E. Daniel St., Champaign, IL 61820; e-mail: gamiller@uiuc.edu.

References

Anderson, N.B., & Scott, P.A. (1999). Making the case for psychophysiology during the era of molecular biology. *Psychophysiology, 36,* 1–14.

Cañive, J.M., Edgar, J.C., Miller, G.A., & Weisend, M.P. (1999, April). *MEG recordings of M300 in controls and schizophrenics.* Paper presented at the biennial meeting of the International Congress on Schizophrenia Research, Santa Fe, NM.

Deldin, P.J., Keller, J., Gergen, J.A., & Miller, G.A. (2000). Right-posterior N200 anomaly in depression. *Journal of Abnormal Psychology, 109,* 116–121.

Fodor, J.A. (1968). *Psychological explanation.* New York: Random House.

Ford, J.M. (1999). Schizophrenia: The broken P300 and beyond. *Psychophysiology, 36,* 667–682.

Hollon, S.D. (1995). Depression and the behavioral high-risk paradigm. In G.A. Miller (Ed.), *The behavioral high-risk paradigm in psychopathology* (pp. 289–302). New York: Springer-Verlag.

Keller, J., Nitschke, J.B., Bhargava, T., Deldin, P.J., Gergen, J.A., Miller, G.A., & Heller, W. (2000). Neuropsychological differentiation of depression and anxiety. *Journal of Abnormal Psychology, 109,* 3–10.

Lang, P.J., Davis, M., & Öhman, A. (in press). Fear and anxiety: Animal models and human cognitive psychophysiology. *Journal of Affective Disorders.*

Marr, D. (1982). *Vision: A computational investigation into the human representation and processing of visual information.* New York: Freeman.

Meany, M.J. (1998, September). *Variations in maternal care and the development of individual differences in neural systems mediating behavioral and endocrine responses to stress.* Address presented at the annual meeting of the Society for Psychophysiological Research, Denver, CO.

Miller, G.A. (1995, October). *How we think about cognition, emotion, and biology in psychopathology:* Presidential address presented at the annual meeting of the Society for Psychophysiological Research, Toronto, Ontario, Canada.

Miller, G.A., & Kozak, M.J. (1993). A philosophy for the study of emotion: Three-systems theory. In N. Birbaumer & A. Öhman (Eds.), *The structure of emotion: Physiological, cognitive and clinical aspects* (pp. 31–47). Seattle, WA: Hogrefe & Huber.

National Institute of Mental Health. (1999, February 12). [Announcement of NIMH workshop, "Emotion and mood"]. Unpublished e-mail.
National Institutes of Health. (1999, February). *Peer review notes*. Washington, DC: Author.
Thase, M.E., Greenhouse, J.B., Frank, E., Reynold, C.F., Pilkonis, P.A., Hurley, K., Grochocinski, V., & Kupfer, D.J. (1997). Treatment of major depression with psychotherapy or psychotherapy-pharmacotherapy combinations. *Archives of General Psychiatry*, 54, 1009–1015.
Zuckerman, M. (1999). *Vulnerability to psychopathology: A biosocial model*. Washington, DC: American Psychological Association.

Critical Thinking Questions

1. To further appreciate the point of this article, consider the following analogy. Suppose you know exactly what your computer does as your friend types a paper using Microsoft Word. You have complete knowledge of when the hard drive is accessed, what happens in the memory registers, and so on. But you have no knowledge of the functional software commands associated with MS Word. In other words, when your friend types command-U, you know what the physical guts of the computer do, but you don't know why your friend types command-U, nor what that means for the document he is producing. Suppose, also, that your friend has a complete, exhaustive knowledge of how to use MS Word, including all of the esoteric features. But your friend does not have the slightest idea of what is happening inside the computer as he uses it. Who understands MS Word better, you or your friend? Is one understanding more important or more fundamental than the other?
2. What would it mean for a psychological process (e.g., long term memory) to be implemented in something other than the brain? What would it mean for a neural circuit *not* to be identified with a single psychological function, but rather with multiple functions?
3. Most of the examples Miller and Keller provide are clinical. Clinical disorders are arguably more high-level than the sort of behaviors that cognitive psychologists study: for example, clinical depression has embedded in it memory, emotion, decision-making and other processes. If cognitive processes are lower level, does that mean that Miller and Keller's arguments do not apply to cognition and the brain?

Automatic Brains—Interpretive Minds

Matthew Roser and Michael S. Gazzaniga
Dartmouth College

Abstract

The involvement of specific brain areas in carrying out specific tasks has been increasingly well documented over the past decade. Many of these processes are highly automatic and take place outside of conscious awareness. Conscious experience, however, seems unitary and must involve integration between distributed processes. This article presents the argument that this integration occurs in a constructive and interpretive manner and that increasingly complex representations emerge from the integration of modular processes. At the highest levels of consciousness, a personal narrative is constructed. This narrative makes sense of the brain's own behavior and may underlie the sense of a unitary self. The challenge for the future is to identify the relationships between patterns of brain activity and conscious awareness and to delineate the neural mechanisms whereby the underlying distributed processes interact.

Keywords

neural correlates of consciousness; interpreter; integration

Although it has been known for more than a century that particular parts of the brain are important for particular functions, the past decade of functional magnetic resonance imaging (fMRI) research has lead to a huge upsurge in evidence for functional specialization. This work has identified areas of the cortex, the convoluted outer layer of the brain, that are involved in processing particular stimulus attributes, or performing certain tasks. For example, cortical areas especially responsive to faces, movement, and places have been found, and these experimental results have been replicated by many independent observers. Although some of the initial claims for functional specialization have been tempered somewhat in the light of new findings, it is becoming ever more clear that the cortex is not a homogeneous, general-purpose computing device, but rather is a complex of circumscribed, modular processes occupying distinct locations.

Most of the work undertaken by these specialist systems occurs automatically and outside of conscious control. For instance, if certain stimuli trick your visual system into constructing an illusion, knowing that you have been tricked does not mean that the illusion disappears. The part of the visual system that produces the illusion is impervious to correction based on such knowledge. Additionally, a convincing illusion can leave behavior unaffected, as when observers are asked to scale the distance between their fingers to the size of a line presented with an arrowhead attached to each end. Although the arrowheads can alter the perceived size of a line (the Müller-Lyer illusion), observers do not

---

Address correspondence to Matthew Roser, Department of Psychological and Brain Sciences, Moore Hall, Dartmouth College, Hanover, NH 03755; e-mail: matthew.rosser@dartmouth.edu.

make a corresponding adjustment in the distance between their fingers, suggesting that the processes determining the overt behavior are isolated from those underlying the perception. Thus, a visuo-motor process in response to a stimulus can proceed independently of the simultaneous perception of that stimulus (Aglioti, DeSouza, & Goodale, 1995).

Stimuli that are not consciously perceived by subjects can, nonetheless, affect behavior. For example, stimuli that are presented very briefly and followed by a masking stimulus go unperceived by subjects, but still activate response mechanisms and speed the recognition of following stimuli that share their semantic properties (Dehaene et al., 1998). When you add to this the observation that robust perceptual aftereffects can be induced by stimuli that are not consciously perceived (Rees, Kreiman, & Koch, 2002), it becomes evident that a great deal of the brain's work occurs outside of conscious awareness and control. Thus, the systems built into our brains carry out their jobs automatically when presented with stimuli within their domain, often without our knowledge.

The most striking evidence for the isolation of function from consciousness comes from studies of patients showing either neglect or blindsight. Neglect is a condition in which the patient ignores a part of space, usually the left; it is typically found in people with damage to the right parietal area of the brain and is thought to be due to the disruption of the brain's mechanisms for allocating attention. Astonishingly, patients with neglect often deny that they have any such condition. It is as if their consciousness of the deficit is destroyed by the lesion just as their actual awareness of a part of space is, even though early visual areas of the brain (i.e., areas that receive and process incoming visual information) are intact and functioning.

The even more bizarre condition known as blindsight describes the residual visual function shown by some patients following a lesion in early visual areas. Although these patients claim to be completely blind in the side of visual space contralateral to the lesion, they are nonetheless able to discriminate, locate, and guide motion toward a stimulus in that area, all without a conscious percept (Rees et al., 2002).

Together, these syndromes and studies in normal subjects suggest that the activity of the brain is not strictly continuous with our conscious experience. Instead, we are sometimes oblivious to complex processing that occurs in the brain. The question then becomes, what determines whether a process is conscious or not?

BRAIN ACTIVITY AND CONSCIOUSNESS

The neural correlates of consciousness in the human brain have been investigated using fMRI and a technique known as binocular rivalry. In this kind of study, a different stimulus is presented to each eye, and the conscious percept typically switches back and forth between the two stimuli, each percept lasting for a few seconds. Subjects indicate when their perception changes from one stimulus to the other, and because the stimuli themselves are static, any changes in neural activity that correlate with a change in the reported percept can be ascribed to changes in the contents of awareness (Tong, Nakayama, Vaughan, & Kanwisher,

1998). Brain activations elicited by rivalrous stimuli are very similar in magnitude and location to activations seen in response to separate stimuli that are presented alternately, suggesting that areas involved in processing a type of stimulus are also involved in the conscious perception of that type of stimulus (Zeki, 2003).

fMRI studies have also revealed substantial brain activations in response to stimuli that are not consciously perceived by subjects (Moutoussis & Zeki, 2002). For example, when color-reversed faces (e.g., an outlined red face on a green background and an outlined green face on a red background) are displayed separately to the two eyes, binocular fusion occurs, and subjects report seeing only the color that results from the combination of the two stimulus color (in this case, yellow). The color inputs to the brain are "mixed," like paint on an artist's palette, and the face stimuli become invisible. Despite not being consciously seen, these stimuli typically activate those areas of the brain that are activated by perceived faces. Why then are some seen and some not?

Brain activations correlated with perceived stimuli and those correlated with unseen stimuli show differences in both their intensity and their spatial extent (Dehaene et al., 2001). Dehaene and his colleagues found that although unperceived stimuli and perceived stimuli activated similar locations in the brain, the activations associated with perceived stimuli were many times more intense than those seen with unperceived stimuli and were accompanied by activity at additional sites. Thus, consciousness may have a graded relationship to brain activity, or a threshold may exist, above which activation reaches consciousness (Rees et al., 2002). At present this issue is unresolved, but the development of increasingly sophisticated designs in fMRI may yield progress by allowing the degree of activation to be determined as the availability of a stimulus to awareness is manipulated.

The increased spatial extent of activations elicited by perceived stimuli in the experiment by Dehaene and his colleagues suggests another possible mechanism for determining whether a stimulus reaches consciousness. Processing of a stimulus may reach consciousness if it is integrated into a large-scale system of cortical activity.

CONSCIOUSNESS SEEMS UNITARY

Despite the evidence that processing is distributed around the brain in functionally localized units, and that much processing proceeds outside of awareness, we personally experience consciousness as a unitary whole. How can these observations be resolved?

One possibility is that processes occurring within localized areas and circumscribed domains become available to consciousness only when they are integrated with other domains. Dehaene and Naccache (2001) have hypothesized that there is a *global neuronal work space* in which unconscious modular processes can be integrated in a common network of activation if they receive amplification by an attentional gating system. Attentional amplification leads to increased and prolonged activation and allows processing at one site to affect processing at another. In this way, brain areas involved in perception, action, and emotion can interact with each other and with circuits that can reinstate past states of this work space.

According to this hypothesis, consciousness is the collection of modular processes that are mobilized into a common neuronal work space and integrated in a dynamic fashion. It is a global pattern of activity across the brain, allowing information to be maintained and influence other processes. For instance, consider the task in which subjects are asked to match the distance between their fingers to the size of a Müller-Lyer figure. If a small delay is introduced between the observation of the figure and the reaching response, subjects must rely on their memory of the perceived size when scaling their grip to the size of the figure. Memory involves a consciously maintained representation. In this situation, the illusion does, in fact, affect the subject's motor response (Aglioti et al., 1995).

This model can explain some of the bizarre deficits of consciousness that occur as the result of brain lesions. As processing that does not achieve amplification remains entirely outside of consciousness, a neglect patient may not be aware of his or her deficit because the mechanism linking local processing to global patterns of activation has been disrupted. Thus, a lesion in a specific location may wipe out not merely processing of an attribute, but also the consciousness of the attribute.

Patients with severe cognitive deficits often confabulate wildly in order to produce an explanation of the world that is consistent with their conscious experience. These confabulations include completely denying the existence of a deficit and probably result from interpretations of incomplete information, or a reduced range of conscious experience (Cooney & Gazzaniga, 2003). Wild confabulations that seem untenable to most people, because of conscious access to information that contradicts them, probably seem completely normal to patients to whom only a subset of the elements of consciousness are available for integration.

MIND IS INTERPRETIVE AND CONSTRUCTIVE

The corpus callosum, which connects the two hemispheres, is the largest single fiber tract in the brain. What happens, then, when you cut this pathway for hemispheric communication and isolate the modular systems of the right hemisphere from those in the left? In the so-called split brain, only processes within a hemisphere can be integrated via cortical routes, and only a limited number of processes that can propagate via subcortical routes can be integrated between the hemispheres. Upon introspection, split-brain patients will tell you that they feel pretty normal. And yet, splitting the brain can reveal some of the most striking disconnections between brain processes and awareness. Each hemisphere can be presented with information that remains unknown to the opposite hemisphere.

Experimental designs that exploit this lack of communication have revealed that the left hemisphere tends to interpret what it sees, including the actions of the right hemisphere (Gazzaniga, 2000). For example, suppose two different scenes are presented simultaneously, one to each hemisphere, and the patient is asked to use his or her left hand to choose an appropriate item from an array of pictures of objects that may or may not be typically found within the presented scenes. The left hand is controlled by the right hemisphere, so the patient's left hemisphere, which has no knowledge of what was presented to the right hemisphere, can observe the subsequent actions of the right hemisphere.

If the patient is asked why he or she chose a particular item, the patient's verbal reply will be largely controlled by the left hemisphere, where the brain's primary language centers are located. Studies using this procedure have shown that patients often reply with an interpretation of the action that is congruent with the scene presented to the left hemisphere. Thus, patients resolve one hemisphere's actions with the other hemisphere's perceptions, by producing an explanation that eliminates conflict between the two. Patients' responses in such studies are very similar to the confabulations produced by brain-damaged patients who deny that they have a serious deficit by rationalizing their bizarre behavior (Cooney & Gazzaniga, 2003).

The hypothesis-generating nature of the left hemisphere has also been demonstrated in a nonlinguistic manner. When each hemisphere of a split brain is asked to predict whether a light will appear on the top or the bottom of a computer screen on a series of trials, and to indicate its prediction by pushing one of two buttons with the contralateral hand, the two hemispheres employ radically different strategies. The right hemisphere takes the simple approach and consistently chooses the more probable alternative, thereby maximizing performance. By contrast, the left hemisphere does what neurologically normal subjects do and distributes its responses between the two alternatives according to the probability that each will occur, despite the fact that this is a suboptimal strategy (Wolford, Miller, & Gazzaniga, 2000). It seems that the left hemisphere is driven to hypothesize about the structure of the world even when this is detrimental to performance.

The left-hemisphere interpreter may be responsible for our feeling that our conscious experience is unified. Generation of explanations about our perceptions, memories, and actions, and the relationships among them, leads to the construction of a personal narrative that ties together elements of our conscious experience into a coherent whole. The constructive nature of our consciousness is not apparent to us. The action of an interpretive system becomes observable only when the system can be tricked into making obvious errors by forcing it to work with an impoverished set of inputs, such as in the split brain or in lesion patients. But even in the damaged brain, this system still lets us feel like "us."

CONCLUSIONS AND FUTURE DIRECTIONS

It is becoming increasingly clear that consciousness involves disunited processes that are integrated in a dynamic manner. It is assembled on the fly, as our brains respond to constantly changing inputs, calculate potential courses of action, and execute responses. But it is also constrained by the nature of modular processes that occur without conscious control, and large parts of it can be destroyed, leaving a rump that operates only within its reduced sphere. Progress toward an overarching theory of consciousness will involve putting our picture of the brain back together. Although carving cognition and brain function up at the joints has been a hugely productive approach, future progress must depend on a variety of approaches that integrate disparate and circumscribed processes.

To this end, developing techniques in brain mapping hold much promise. Statistical analysis of fMRI data allows the correlations between activations in different areas to be assessed, yielding maps of cerebral interactivity. The application of these techniques to investigation of the neural correlates of consciousness is extremely relevant, as the activation of large networks is thought to be a necessary condition for consciousness.

A further step involves integrating maps of cerebral interactivity with data about neuroanatomical connections. This technique allows a subset of brain processes to be explicitly modeled as a functional network and yields a map of the strengths of anatomical connections that best fits the imaging data (Horwitz, Tagamets, & McIntosh, 1999). At present, much of the data on neuroanatomical connections comes from postmortem studies in monkeys, but a developing noninvasive MRI technique known as diffusion-tensor imaging (DTI) allows the paths of neurons to be tracked and should provide more accurate data about the human brain. DTI is set to have a huge future impact on this field (Le Bihan et al., 2001).

The brain sciences of the coming years promise to yield great progress in our understanding of integrative processes in the brain. The ultimate aim is to come to a theory of consciousness that, while acknowledging that our brains are elaborate assemblies of myriad processes, explains how it is that we feel so unified.

Recommended Reading

Driver, J., & Mattingley, J.B. (1998). Parietal neglect and visual awareness. *Nature Neuroscience, 1*, 17–22.

Gazzaniga, M.S. (2000). (See References)

Savoy, R.L. (2001). History and future directions of human brain mapping and functional neuroimaging. *Acta Psychologica, 107*, 9–42.

Acknowledgments—Preparation of this article was supported by Grant NS31443 from the National Institutes of Health. We are grateful to Margaret Funnell, Paul Corballis, and Michael Corballis for interesting discussions on the topics covered here.

References

Aglioti, S., DeSouza, J.F., & Goodale, M.A. (1995). Size-contrast illusions deceive the eye but not the hand. *Current Biology, 5*, 679–685.

Cooney, J.W., & Gazzaniga, M. (2003). Neurological disorders and the structure of human consciousness. *Trends in Cognitive Sciences, 7*, 161–165.

Dehaene, S., & Naccache, L. (2001). Towards a cognitive neuroscience of consciousness: Basic evidence and a workspace framework. *Cognition, 79*, 1–37.

Dehaene, S., Naccache, L., Cohen, L., Bihan, D.L., Mangin, J.F., Poline, J.B., & Riviere, D. (2001). Cerebral mechanisms of word masking and unconscious repetition priming. *Nature Neuroscience, 4*, 752–758.

Dehaene, S., Naccache, L., Le Clec, H.G., Koechlin, E., Mueller, M., Dehaene-Lambertz, G., van de Moortele, P.F., & Le Bihan, D. (1998). Imaging unconscious semantic priming. *Nature, 395*, 597–600.

Gazzaniga, M.S. (2000). Cerebral specialization and interhemispheric communication: Does the corpus callosum enable the human condition? *Brain, 123*, 1293–1326.

Horwitz, B., Tagamets, M.-A., & McIntosh, A.R. (1999). Neural modeling, functional brain imaging, and cognition. *Trends in Cognitive Sciences, 3*, 91–98.

Le Bihan, D., Mangin, J.F., Poupon, C., Clark, C.A., Pappata, S., Molko, N., & Chabriat, H. (2001). Diffusion tensor imaging: Concepts and applications. *Journal of Magnetic Resonance Imaging, 13*, 534–546.

Moutoussis, K., & Zeki, S. (2002). The relationship between cortical activation and perception investigated with invisible stimuli. *Proceedings of the National Academy of Sciences, USA, 99*, 9527–9532.

Rees, G., Kreiman, G., & Koch, C. (2002). Neural correlates of consciousness in humans. *Nature Reviews Neuroscience, 3*, 261–270.

Tong, F., Nakayama, K., Vaughan, J.T., & Kanwisher, N. (1998). Binocular rivalry and visual awareness in human extrastriate cortex. *Neuron, 21*, 753–759.

Wolford, G., Miller, M.B., & Gazzaniga, M. (2000). The left hemisphere's role in hypothesis formation. *Journal of Neuroscience, 20*, RC64.

Zeki, S. (2003). The disunity of consciousness. *Trends in Cognitive Sciences, 7*, 214–218.

Critical Thinking Questions

1. Why do you think the human brain is set up to generate hypotheses about the world? Why don't we simply accept the world as it is, and say "I don't know" when there are actions we don't understand?
2. The left hemisphere seems to play a significant role in hypothesis generation. It is tempting to propose that this laterality is related to the left-hemisphere dominance for language. Do you think that babies who don't yet have a command of language also generate hypotheses? Do you think that animals do?
3. What does Roser and Gazzaniga's theory say about the anatomic location of consciousness?

The Mentality of Apes Revisited

Daniel J. Povinelli[1] and Jesse M. Bering

Cognitive Evolution Group, University of Louisiana, Lafayette, Louisiana (D.J.P.), and Department of Psychology, Florida Atlantic University, Boca Raton, Florida (J.M.B.)

Abstract

Although early comparative psychology was seriously marred by claims of our species' supremacy, the residual backlash against these archaic evolutionary views is still being felt, even though our understanding of evolutionary biology is now sufficiently advanced to grapple with possible cognitive specializations that our species does not share with closely related species. The overzealous efforts to dismantle arguments of human uniqueness have only served to show that most comparative psychologists working with apes have yet to set aside the antiquated evolutionary "ladder." Instead, they have only attempted to pull chimpanzees up to the ladder's highest imaginary rung—or perhaps, to pull humans down to an equally imaginary rung at the height of the apes. A true comparative science of animal minds, however, will recognize the complex diversity of the animal kingdom, and will thus view *Homo sapiens* as one more species with a unique set of adaptive skills crying out to be identified and understood.

Keywords

chimpanzees; cognitive evolution; theory of mind; comparative psychology

Five to seven million years ago, a small lineage of anthropoid apes came down from the trees. Within a couple of million years, descendants of this lineage had evolved a new form of locomotion (striding bipedalism), and had resculpted their pelvic girdle, head, hands, and feet. They tripled the size of their brain, and even appeared to have reorganized some of the most fundamental systems within that brain (see Preuss & Coleman, in press). In a world already teeming with biological diversity, the human lineage made its debut.

With the appearance of our species came the ability to ponder those origins, and to pose such questions as, what does it mean to be human? and, more central to this essay, what psychological characteristics appear to be uniquely human? Such questions have challenged generations of inquisitive minds, all the while fueling controversy and divisiveness. Typically, the answers to such questions depended on the profession of the individuals being asked: To the theologian, the uniquely human endowment was the possession of a soul; to the psychologist, it was language; to the anthropologist, it was culture.

DEMOLISHING HUMAN UNIQUENESS

Alas, enter the first comparative psychologist, Darwin, who, running against centuries of religious and philosophical dogma, strategically announced that there is no characteristic truly unique to humans. "There can be no doubt," wrote Darwin in 1871, "that the difference between the mind of the lowest man and that of the highest animal is immense. Nevertheless the difference, great as it is, certainly is one of degree and not of kind" (Darwin, 1871/1982, p. 445). How could Darwin be so sure? To him and his followers, the answer was simple: Just observe other species' natural, spontaneous behaviors, and then use introspection to infer the underlying causes of these behaviors. Although this may seem like a sensible enough approach, consider the full implications of this method: It makes the human mind the standard against which all other minds are judged, installing our mental processes—and only ours—into the minds of other species. Even now, as data calling for a radical departure from this canonical view continue to mount, this most anthropomorphic (and ultimately un-evolutionary) of assumptions continues to live on in the field of comparative psychology.

In no case is this truer than in research into the mental abilities of chimpanzees and other great apes. Here, the classic argument by analogy enjoys the protection of the suspect notion of "evolutionary plausibility." Researchers regularly assert that the parsimonious explanation of behavioral similarity between humans and chimpanzees is the operation of equally similar psychological systems. Against this theoretical backdrop, Savage-Rumbaugh, mentor of the bonobo chimpanzee Kanzi, writes a book whose subtitle proclaims that her chimpanzee is "the ape at the brink of the human mind" (Savage-Rumbaugh & Lewin, 1994). In a recent article, Suddendorf and Whiten (2001) conclude that "the gap between human and animal mind has been narrowed" (p. 644). And de Waal's (1999) take on the same trend is that the chimpanzee is "inching closer to humanity" (p. 635).

For some researchers, continuity extends to identity. Savage-Rumbaugh, for example, declares that she has met the mind of another species and discovered that it is human: "I found out that it was the same as ours," she concludes. "I found out that 'it' was me!" (as quoted by Dreifus, 1999, p. 54). More typically, though, chimpanzees are caricatured as watered-down human beings. Echoing Darwin, Fouts (1997) sees any attempt to demonstrate differences between closely related species as symptomatic of "Cartesian delusions," and proclaims, "The cognitive and emotional lives of animals differ only by degree, from the fishes to the birds to monkeys to humans" (p. 372). Likewise, Goodall (1990) writes of a "succession of experiments that, taken together, clearly prove that many intellectual abilities that had been thought unique to humans were actually present, though in a less highly developed form, in other, non-human beings" (p. 18).

All of this adds up to an agenda for psychological research with chimpanzees: "Just how human are chimpanzees?" We suggest, however, that the obsession with establishing psychological continuity between humans and other apes has cast this area of comparative psychology into a great freeze. It has contributed to marginalizing the discipline's mission by reducing it to a series of demonstrations in which one humanlike ability after another is revealed in nonhuman animals. It is an objective anchored to the mistaken idea that evolution proceeds linearly and that apes are thus playing catch-up to the human intellect. This objective, however, is fundamentally at odds with the central theme of modern biology: Evolution is real, and it produces diversity.

Indeed, differences are seen as somehow obscuring the true evolutionary relationships among living species: "Researchers are regularly finding heretofore unexpected realms and degrees of similarity," noted Russon and Bard (1996), "and these similarities are particularly useful for evolutionary reconstructions" (p. 14). Not only is this point of view 180° out of phase with modern cladistic approaches to evolutionary reconstruction, but if the dramatic resculpting of the human body and brain that occurred over the past 4 million years or so involved the evolution of some qualitatively new cognitive systems, then this insistence on focusing on similarities will leave comparative psychologists unable to investigate hallmarks of their own species—or chimpanzees, for that matter. It is an agenda that does justice to no one.

AN ALTERNATIVE FRAMEWORK: THE REINTERPRETATION HYPOTHESIS

Perhaps the greatest obstacle to overcoming Darwin's a priori straitjacket of unbroken psychological continuity has been the difficulty of imagining an alternative. After all, if there really were a viable alternative, surely whatever intuitive anthropomorphic biases researchers have could be overcome—much in the way that Newtonian mechanics overcame tenets of Aristotelian physics. But here the challenge may be more substantial, because in this case, the very system that comparative psychologists seek to investigate is the one producing the illusion, compelling these researchers to recreate the psychology of other species in their own image.

In recent publications, we have suggested an alternative to the continuity paradigm, an alternative that we initially applied to the evolution of a *theory of mind*—the ability to reason about mental states in the self and others (for an elaboration of this model, see Povinelli, 2000, chap. 2). Our alternative posits that for dozens of millions of years, the primate order produced numerous social species, each one inheriting a core stock of mammalian social behaviors and then tweaking these behaviors to cope with the peculiar demands of its own circumstances. Although there is debate as to the key factors in this process, there can be no doubt that natural selection for social living was intense during the radiation of the primates, and with this drive to sociality came selection for social behavior to both exploit and cope with group living.

However, rather than positing sociality as the prime mover for the evolution of psychological systems for representing other minds, our alternative holds that the vast array of spontaneous behaviors that humans share with chimpanzees, including deception, gaze following, holding a grudge, tool use, reconciliation, and organized hunting, emerged and were in full operation long before additional systems evolved to interpret these behaviors in terms of underlying mental states. Instead, these behaviors were generated through existing psychological processes, motivated by physiological, attentional, perceptual, and affective mechanisms—mechanisms that continue to guide an enormously complicated assemblage of primate (including human) behavior.

The final part of our claim is that it was not until a particular lineage appeared—the human one—that a new representational system was stamped into the old, so that the observable world, and those things transpiring in it, were "reinterpreted" with hidden meaning, allowing humans to reflect on unobservable causes, such as mental states. Without discarding the ancestral mechanisms it built on, this novel causal explanatory system then generated its own subassemblage of behaviors (e.g., progressive cultural transmission, religious rituals, and explicit pedagogy), all of which hinge on the ability to represent a social and physical world governed by abstract causal forces. Because it envisions that a hallmark of human mental evolution was installing a system or series of systems that interprets ancestral behaviors in new ways, we have labeled our model the *reinterpretation hypothesis*.

The reinterpretation hypothesis calls into question the fairly common assertion that if similar behaviors "are the product of a common history, then it is likely that the underlying psychological processes responsible for the overt behavior are similar, too" (Suddendorf & Whiten, 2001, p. 643). To the contrary, the reinterpretation hypothesis makes clear that no a priori argument from similarities in spontaneous behavior will suffice. Although the recently shared ancestral heritage of chimpanzees and humans virtually guarantees behavioral homologies, the totality of the representational software that rides alongside (or in some cases causes) similar behaviors in the two species is not necessarily the same. With this alternative framework—a theoretical approach that embraces both similarity and difference—it is now possible to return to the investigation of human uniqueness in the way that a biologist would address an investigation of the specializations of any species—open to wherever the empirical facts seem to lead.

DIFFERENCES IN THE MENTALITIES OF APES AND HUMANS

Evidence that human evolution was marked by the emergence of novel mental abilities is beginning to accumulate. There is increasing evidence that, at some point after hominids separated from the line leading to the modern African apes, humans developed a unique capacity to mentally represent a world of hidden causal forces, including mental states. Consider the following:

- Although chimpanzees respond to eye gaze by following the visual trajectory of other individuals, even around barriers, they do not appear to grasp the fact that others' visual behaviors are accompanied by the psychological experience of "seeing" (Fig. 1; see the review by Povinelli, 2000). Recent widely reported claims that chimpanzees may attribute the mental states of seeing and knowing to other chimpanzees (e.g., Hare, Call, Agnetta, & Tomasello, 2000) have not been supported by attempts at replication (Karin-D'Arcy & Povinelli, 2002).
- In carefully controlled studies, great apes have failed to appreciate the underlying referential nature of intentional communication (e.g., Povinelli, Reaux, Bierschwale, Allain, & Simon, 1997). Whether the communicative attempt comes in the form of an extended index finger (i.e., pointing) or an iconic device (e.g., a replica of a box containing a food reward), chimpanzees do not seem to understand that the communicative behaviors of other individuals are driven by a desire to share information.
- A number of nonverbal methodological attempts to parallel research on children's understanding of the mental state of belief have shown that chimpanzees do not distinguish between individuals who are ignorant versus knowledgeable. For example, they respond to observers who have

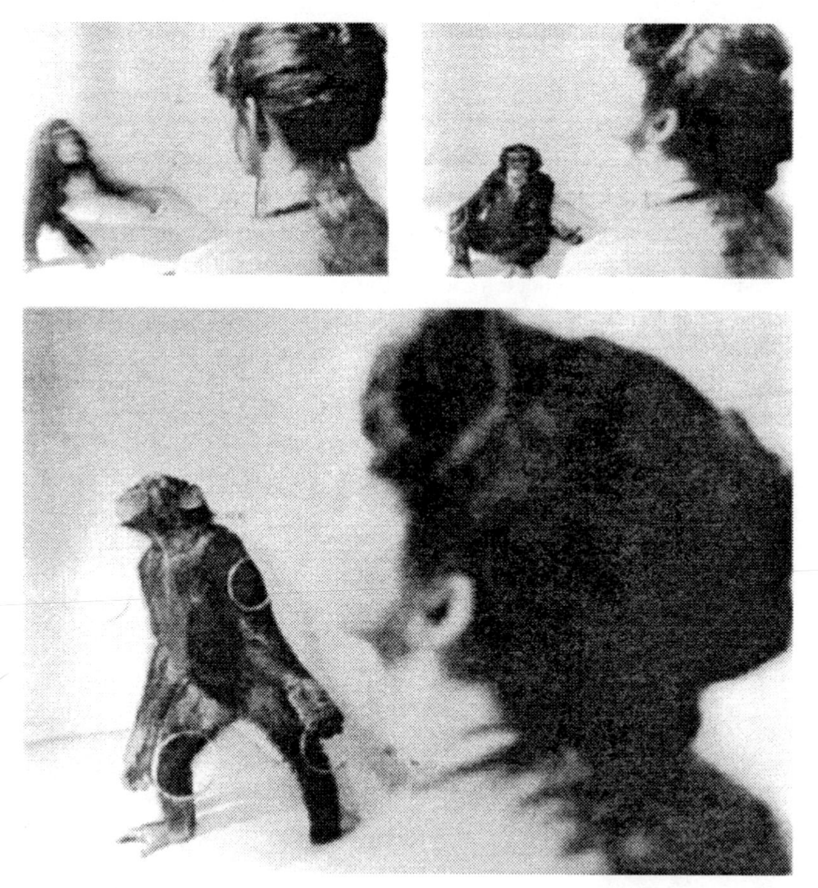

Fig. 1. Gaze following in a 6-year-old chimpanzee. The chimpanzee makes eye contact with a human caretaker (top left), who turns and looks above the chimpanzee (top right); the chimpanzee then follows the caretaker's gaze (bottom). Although it is tempting to assume in such circumstances that the chimpanzee's interpretation of the situation is similar to our own (i.e., that the animal can infer that the person has "seen" something), the reinterpretation hypothesis (see the text for details) makes clear that in this and many other cases, high-level human cognitive systems may have been grafted into a suite of ancient systems that modulate quite ancient behavioral patterns. Only programmatic experimental approaches will suffice to determine the presence or absence of such systems in other species.

witnessed the hiding of food no differently than they do to those who are oblivious to its actual location—choosing at random who they would like to retrieve the reward for them (Call & Tomasello, 1999).

- "Expert" chimpanzees that have been previously trained on how to perform a cooperative task (e.g., jointly pulling a heavy box by two separate ropes) with a human partner do not spontaneously guide naive chimpanzee partners on relevant dimensions of the task; these experts essentially ignore the fact that their new partners lack the requisite knowledge for success, and fail to instruct them through teaching behaviors (e.g., showing, touching, pointing; see Povinelli & O'Neill, 2000).

Interestingly, this pattern may translate to the nonsocial domain as well. Recently, we completed a project that was designed to map our chimpanzees' understanding of unobservable forces in the physical world (see Povinelli, 2000). The initial round of nearly 30 studies, conducted over a 5-year period, was centered on the widely celebrated ability of chimpanzees to make and use simple tools. However, we were interested not in the level of complexity that such tool use and construction might achieve, but rather whether chimpanzees reason about the hidden properties and functions of tools. In particular, we asked whether chimpanzees' understanding of the physical world is mediated by concepts robustly in place by about 3 years of age in human children—things such as gravity, force, shape, physical connection, and mass.

The results converged upon a finding strikingly analogous to what we have described about chimpanzees' understanding of the social world: Although they are very good at understanding and learning about the observable properties of objects, they appear to have little or no understanding that these observable regularities can be accounted for, or explained, in terms of unobservable causal forces. In short, we have speculated that for every unobservable causal concept that humans may form, the chimpanzee will rely exclusively upon an analogue concept, constructed from the perceptual invariants that are readily detectable by the sensory systems.

CONCLUDING REMARKS

If there is one thing that our species is obviously not very good at, it is imagining ways of understanding the world that differ markedly from our own. The popular press overflows with stories of empathic gorillas rescuing young children, cats scaling burning buildings to bring their kittens to safety, and dogs who think they are human. Whatever the behavioral facts of these cases, one thing is certain: We humans will automatically interpret the psychological facts from the perspective of our evolved, but peculiarly distorted, ways of understanding the world.

Research with chimpanzees and other great apes remains marginalized within the cognitive and biological sciences largely because the field has failed to come to grips with the most important tenet of modern biology: Evolution is real, and it produces diversity. Comparative psychology was founded upon the notion that organisms could be arranged into an evolutionary scale or ladder in which the mental operations of living species were said to differ in degree, not kind. Most psychologists who work with chimpanzees continue to espouse this view. Because of the importance of comparing and contrasting the psychological systems of our own species with those of our nearest living relatives, there is an overwhelming need to train the next generation of these psychologists in the intricacies of the evolutionary biology of the organisms they study. Organismal biology can provide the theoretical motivation to look for, and thus celebrate, the marvelous psychological differences that exist among species.

Recommended Reading

Bering, J.M. (in press). The existential theory of mind. *Review of General Psychology*.

Hodos, W., & Campbell, C.B.G. (1969). *Scala naturae*: Why there is no theory in comparative psychology. *Psychological Review*, 76, 337–350.

Kohler, W. (1927). *The mentality of apes* (2nd ed.). New York: Vintage Books.
Povinelli, D.J. (2000). (See References)
Tomasello, M. (1999). *The cultural origins of human cognition.* Cambridge, MA: Harvard University Press.

Acknowledgments—This work was supported by a Centennial Fellowship Award to D.J.P. from the James S. McDonnell Foundation.

Note

1. Address correspondence to Daniel Povinelli, Cognitive Evolution Group, University of Louisiana, 4401 W. Admiral Doyle Dr., New Iberia, LA 70560; e-mail: ceg@louisiana.edu.

References

Call, J., & Tomasello, M. (1999). A nonverbal theory of mind test: The performance of children and apes. *Child Development, 70*, 381–395.
Darwin, C. (1982). *The descent of man.* New York: Modern Library. (Original work published 1871)
de Waal, F.B.M. (1999). Animal behaviour: Cultural primatology comes of age. *Nature, 399*, 635.
Dreifus, C. (1999, August-September). Going ape. *Ms., 9*(5), 48–54.
Fouts, R. (1997). *Next of kin.* New York: William Morrow and Co.
Goodall, J. (1990). *Through a window.* Boston: Houghton Mifflin.
Hare, B., Call, J., Agnetta, B., & Tomasello, M. (2000). Chimpanzees know what conspecifics do and do not see. *Animal Behaviour, 59*, 771–785.
Karin-D'Arcy, R., & Povinelli, D.J. (2002). *Do chimpanzees know what each other see? A closer look.* Manuscript submitted for publication.
Povinelli, D.J. (2000). *Folk physics for apes.* New York: Oxford University Press.
Povinelli, D.J., & O'Neill, D.K. (2000). Do chimpanzees use their gestures to instruct each other? In S. Baron-Cohen, H. Tager-Flusberg, & D.J. Cohen (Eds.), *Understanding other minds* (pp. 459–487). Oxford, England: Oxford University Press.
Povinelli, D.J., Reaux, J.E., Bierschwale, D.T., Allain, A.D., & Simon, B.B. (1997). Exploitation of pointing as a referential gesture in young children, but not adolescent chimpanzees. *Cognitive Development, 12*, 423–461.
Preuss, T.M., & Coleman, G.Q. (in press). Human-specific organization of primary visual cortex: Alternating compartments of dense Cat-301 and calbindin immuno-reactivity in layer 4A. *Cerebral Cortex.*
Russon, A., & Bard, K. (1996). Exploring the minds of the great apes: Issues and controversies. In A.E. Russon, K.A. Bard, & S.T. Parker (Eds.), *Reaching into thought* (pp. 1–20). Cambridge, England: Cambridge University Press.
Savage-Rumbaugh, S., & Lewin, R. (1994). *Kanzi: The ape at the brink of the human mind.* New York: John Wiley & Sons.
Suddendorf, T., & Whiten, A. (2001). Mental evolution and development: Evidence for secondary representation in children, great apes and other animals. *Psychological Bulletin, 127*, 629–650.

Critical Thinking Questions

1. If you believe that animals and humans are on a continuum of cognitive ability, with different animals varying in how far they are below humans on the "ladder," then it is easy to understand why it would be interesting to study the cognitive ability of animals. Povinelli and Bering specifically say that it is wrongheaded to think of such a continuum. If so, why should we study the cognitive abilities of animals?

2. One reason it is so tempting to think of such a continuum is that animals are so obviously worse than humans in many cognitive abilities that we deem important, for example communication and complex problem-solving. If Povinelli and Bering are correct in that animals are not less cognitively able than we are, but rather just have different cognitive abilities than we do, then some animals should have cognitive abilities that far outstrip ours. Can you think of some examples? Why might examples be difficult to think of, given the world in which we live?

3. Domestic dogs are better than great apes in understanding human communicative signals indicating the location of hidden food (Hare et al., 2002), which is one measure of theory-of-mind. It is thought that this ability is the product of domestication and selective breeding. (Wolves don't show this ability.) Would Povinelli and Bering say that this result supports their thesis? Why or why not?

Hare, B., Brown, M., Williamson, C., & Tomasello, M. (2002). The domestication of social cognition in dogs. *Science, 298,* 1634-1636.

The Projective Way of Knowing: A Useful Heuristic That Sometimes Misleads

Raymond S. Nickerson[1]

Psychology Department, Tufts University, Medford, Massachusetts

Abstract

For many purposes, people need a reasonably good idea of what other people know. This article presents an argument and considers evidence that people use their own knowledge as a basis for developing models of what specific other people know—in particular, that they tend to assume that other people know what they know. This is a generally useful heuristic, but the assumption is often made uncritically, with the consequence that people end up assuming that others have knowledge that they do not have.

Keywords

knowledge; projection; false consensus; expertise; egocentrism

People's behavior is influenced in many ways by what they know about what other people know. Effective conversation, for example, depends not only on shared knowledge between participants, but also on each person having knowledge, or making reasonably accurate assumptions, about what the other knows.

BUILDING A CONCEPTUAL MODEL OF WHAT ANOTHER PERSON KNOWS

Over time, one can develop a detailed conceptual model of what a specific other person (spouse, sibling, friend, associate) knows, fine-tuning and updating the model with information gleaned from frequent interactions. But what does one use for a model of what a stranger knows? How does one cope with the task of communicating with a collection of people—an audience to whom one has to give a talk, or the readership of a newspaper for which one is writing an article—when one has few specifics about its composition? I assume that the basis for the construction of a default model of what a random other person knows is one's model of what one knows oneself.

What an individual knows changes over time. It follows that if a model of another person's knowledge is to be and remain functionally accurate, it too must change on a continuing basis. Several researchers have noted that refining one's model of another person's knowledge dynamically is important if communication is to be successful.

These ideas are incorporated in Figure 1, a conceptualization of how an individual develops a model of another person's knowledge (from Nickerson, 1999). According to this conceptualization, one's model of one's own knowledge serves as a default model of what a random other person knows. This default model is transformed, as individuating information is acquired, into models of specific other individuals. The models of specific others are continually refined and updated as

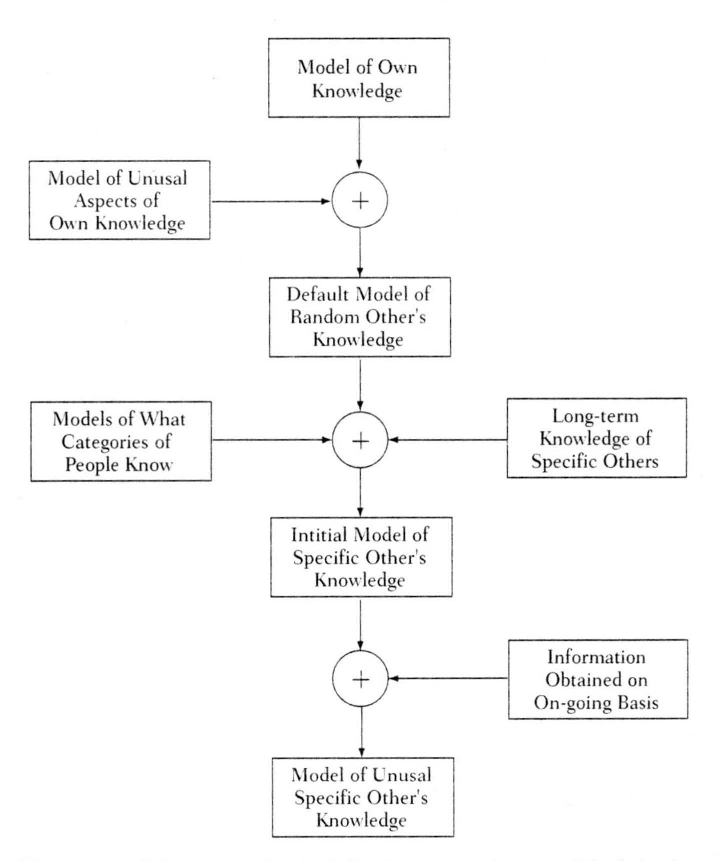

Fig. 1. Illustration of how an individual develops a working model of the knowledge another person has (from Nickerson, 1999).

new information that is relevant to them is acquired. This article focuses on the idea that what serves as the point of departure for developing a model of another person's knowledge is what one knows, or thinks one knows, oneself.

SELF AS A SOURCE OF HYPOTHESES ABOUT WHAT OTHERS KNOW

The notion that a basic source of assumptions or hypotheses regarding what a random other person knows is what one knows oneself has some currency among psychologists. It is closely related to the *simulation* view of how one individual understands another, according to which one imagines oneself in the other's place and discovers how one would think or feel in that situation (Gordon, 1986).

The Projective Way of Knowing

The idea that we understand others by assuming that they are like ourselves is intuitively compelling, and there is much evidence to support the notion that people employ this "projective mode of knowing," as O'Mahony (1984, p. 58) has called it. People who engage in a particular behavior estimate that behavior to be more prevalent than do people who do not engage in that behavior. When attempting to assess the attitudes of specified groups, people tend to project their own attitudes onto those groups. People who are experiencing an experimentally induced emotional state are more likely to project that state to others than are people who are not experiencing it. People generally see their own attitudes and behavior as rational or normative, and they see attitudes and behavior that differ greatly from their own as irrational or deviant. In the political arena, extremists on both ends of the left-right continuum tend to doubt the rationality of those on the opposite end. People sometimes take their own behavior as the norm even in the light of sample-based information to the contrary.

Advantages of Using the Self as a Basis for Developing a Model

Our own knowledge of how we would behave or react in specific situations can be a useful basis, arguably the best basis we have, for anticipating how other people will behave or react in those situations (Hoch, 1987). Projecting our own feelings and reactions to others works because, in fact, people do react similarly in specific situations. The idea that we do well to assume that others are like ourselves is captured in the *principle of humanity*, according to which when trying to understand what someone has said, especially something ambiguous, we should impute to the speaker beliefs and desires similar to our own.

Using our own knowledge as a default model of what a random other person knows simplifies life. If we could not assume, in the absence of contrary evidence, that other people are much like ourselves, the problem of communicating effectively would be overwhelmingly difficult (Davidson, 1982).

THE RISK OF OVERIMPUTATION OF OUR OWN KNOWLEDGE

Although, in the absence of more direct information, one's model of one's own knowledge may be as good a basis as there is for a default assumption about what a random other person knows, evidence suggests that the tendency to impute our own knowledge to others often leads us to assume that others have knowledge they do not actually have, and this can impede communication and mutual understanding in various ways.

Failure to make sufficient allowance for the difference between one's own subjective experiences or perspectives and those of one's hearers or readers has been noted as a source of difficulties in communication in both spoken and written form. Teaching can be adversely affected if teachers underestimate the difference between their own knowledge and that of their students; products of technology may be designed suboptimally if designers underestimate how much difficulty people other than themselves will have in learning to use them. Piaget (1962) remarked on the difficulty that beginning instructors have in placing themselves in the shoes of students who do not know what they themselves do about a course's subject matter, and surmised that they are likely to give incomprehensible lectures for a while as a consequence. Flavell (1977) similarly pointed out that one's own viewpoint can work as an impediment to attaining an accurate appreciation of the viewpoint of another person, and that it may be especially difficult for one to appreciate fully the ignorance of another person with respect to something one understands very well oneself.

The False-Consensus Effect

The *false-consensus effect* refers to a tendency to see oneself as more representative of other people than one really is. The results of numerous studies suggest that the tendency is very common and that it manifests itself in a variety of ways. People are likely to overestimate the amount of general consensus on beliefs and opinions they themselves hold, and to underestimate the degree of agreement on beliefs and opinions that differ from theirs. The effect is illustrated by the finding that U.S. voters typically overestimate the popularity of their favored candidate in a presidential election (Brown, 1982), as well as the extent to which the positions of favored candidates correspond to their own (Page & Jones, 1979).

The Curse of Expertise

People who are experts in specific areas and who recognize themselves as such must realize, almost by definition, that they know more than most other people with respect to their areas of expertise. Nevertheless, the results of several studies suggest that although experts assume that others know less than they about their areas of expertise, they may still overestimate what others know.

The point is illustrated by a study by Hinds (1999), who found that experts in performing a task were more likely than people with only an intermediate level of expertise to underestimate the time novices would take to complete the task. Experts also proved to be resistant to debiasing techniques intended to reduce the tendency to underestimate how difficult novices would find a task to be. In laboratory studies, participants who have been given privileged information for purposes of an experiment may behave as though other participants also have that information even when, if asked, they acknowledge that the other participants do not have the information.

Egocentric Bias in Imputing General Knowledge

The results of many studies suggest that people's estimates of what general knowledge other people have tend to be biased in the direction of the knowledge they themselves have or think they have. When college students were asked to answer general-knowledge questions and to estimate, for each question, the percentage of other college students who would be able to answer that question correctly, they gave higher estimates for questions they thought they knew the answers to (as indicated by confidence ratings), even when their own answers were wrong, than for questions they knew they did not know the answers to, and they were more likely to overestimate the commonality of knowledge if they themselves had it than if they did not.

When students living in New York City rated their familiarity with each of 22 landmarks in the city and estimated the proportions of other city residents who would be able to identify them, students who were highly confident of being able to identify specific landmarks judged those landmarks to be more familiar to others than did students who were not very confident of their own ability to identify them. Students who could identify pictured public figures by name rated the individuals as more recognizable than did students who could not identify them by name. Students who could identify an everyday object estimated the proportion of peers who would be able to identify that object to be higher than did people who could not identify it.

When instructors attempted to answer quiz questions as they expected their students would answer them, they provided about twice as many correct answers as did their students on average. Readers of the account of a conversation attributed to the listener the same understanding of the speaker's utterance as their (the readers') own, even when the utterance was ambiguous and the readers knew that the listener did not have the disambiguating information they had. When observers judged the likelihood that listeners would believe a message that contradicted the listeners' prior belief about a situation, they judged the likelihood to be higher if they (the observers) knew the message to be true than if they knew the message to be false, even though they were aware that the listeners could not have the basis they (the observers) had for judging the message to be true.[2]

A particularly striking example of overimputing one's own knowledge to others comes from an experiment by Newton (1990) in which some participants tapped the rhythms of well-known songs and others attempted to identify the songs on the basis of the tapped rhythms. Tappers estimated the likelihood that listeners would be able to identify the songs to be about .5; the actual probability of correct identification was about .025. Apparently the tappers, who could imagine a musical rendition of a song when they tapped its rhythm, found it overly easy to project their own subjective experience to the listeners, who did not share it.

Illusion of Simplicity

The *illusion of simplicity* refers to the mistaken impression that something is simple just because one is familiar with it. It is illustrated by the findings that people are likely to judge anagrams to be easier to solve if they have been shown the solutions than if they have not and that they are likely to judge sentences to be appropriate for a lower-grade reading level if they have read them before than if they have not (Kelley, 1999). It is a short step from perceiving something to be simpler than it is, only because one is familiar with it, to assuming that someone else, who is not familiar with it, will perceive it the same way.

ANCHORING AND INADEQUATE ADJUSTMENT

This conceptualization of how we build models of what others know can be seen as a case of the general reasoning heuristic of *anchoring and adjustment* (Tversky & Kahneman, 1974), according to which people make judgments by starting with an "anchor" as a point of departure and then make adjustments to it. In this case, the anchor for one's default model of what someone else knows is one's model of what one knows oneself.

Many of the experimental results noted in this article support the idea that, in serving as the anchor for one's model of another person's knowledge, one's model of one's own knowledge is adjusted to take into account differentiating information either about one's own knowledge or about the other person's knowledge. However, as in other documented instances of anchoring and adjustment, the adjustment is often not as great as it should be, and one ends up assuming that another person has knowledge that he or she does not have.

CONCLUDING COMMENTS

In this article, I have emphasized the prevalence of the tendency to overimpute one's own knowledge to other people and the fact that this can be problematic in several respects. Dawes (1989) has made the point that it is possible to err in the other direction as well, and has argued that an uncritical assumption of dissimilarity between oneself and others can also have undesirable consequences. This seems an important caution to bear in mind in interpreting the results noted here. Presumably, relatively accurate models of what others know are more useful than models that are biased either toward or away from what one knows oneself; however, on balance, the literature suggests that biasing one's model of another person's knowledge in the direction of one's own knowledge is a more common problem than biasing it in the opposite direction.

What can be done to improve our conceptions of what specific other people know? I have suggested several possibilities elsewhere (Nickerson, 1999). Here, I mention only the belief that this problem, like many others relating to cognitive or judgmental biases, stems, at least in part, from a failure to be very reflective about assumptions we make—from failing to give much attention to alternative assumptions that could be made. If we generally tend to assume that a random other person knows a fact that we know ourselves, and if we give insufficient consideration to reasons why the other person might not know that fact, we are likely to overimpute our own knowledge to others as a rule.

In many cases, failure to be more critical of our own assumptions may be defended on the grounds that, although the conclusions that we settle on may not be optimal, they are usually close enough for practical purposes and finding better ones would not be worth the effort. However, if judgments of a particular type are relatively consistently biased in a specified way, as judgments of what others know appear to be, search for effective debiasing techniques seems warranted. Simple awareness of a tendency to overimpute one's own knowledge to others may be helpful, but probably not fully corrective. How best to teach people to make more accurate estimates of what other people know, and to counteract the tendency to overimpute their own knowledge to others, remains a challenge to research.

Recommended Reading

Camerer, C., Loewenstein, G., & Weber, M. (1989). The curse of knowledge in economic settings: An experimental analysis. *Journal of Political Economy*, 97, 1232–1254.

Cronbach, L. (1955). Processes affecting scores on "understanding others" and "assumed similarity." *Psychological Bulletin*, 52, 177–193.

Krauss, R., & Fussell, S. (1991). Perspective-taking in communication: Representations of others' knowledge in reference. *Social Cognition, 9*, 2–24.

Mullen, B. (1983). Egocentric bias in estimates of consensus. *Journal of Social Psychology, 121*, 31–38.

Srull, T., & Gaelick, L. (1983). General principles and individual differences in the self as a habitual reference point: An examination of self-other judgments of similarity. *Social Cognition, 2*, 108–121.

Notes

1. Address correspondence to Raymond S. Nickerson, 5 Gleason Rd., Bedford, MA 01730; e-mail: r.nickerson@tufts.edu.

2. References for all the studies alluded to in the preceding three paragraphs are listed in Nickerson (1999).

References

Brown, C. (1982). A false consensus bias in 1980 presidential preferences. *Journal of Social Psychology, 118*, 137–138.

Davidson, D. (1982). Paradoxes of irrationality. In R. Wollheim & J. Hopkins (Eds.), *Philosophical essays on Freud* (pp. 289–305). Cambridge, England: Cambridge University Press.

Dawes, R. (1989). Statistical criteria for establishing a truly false consensus effect. *Journal of Experimental Social Psychology, 25*, 1–17.

Flavell, J. (1977). *Cognitive development*. Englewood Cliffs, NJ: Prentice-Hall.

Gordon, R. (1986). Folk psychology as simulation. *Mind and Language, 1*, 158–171.

Hinds, P. (1999). The curse of expertise: The effects of expertise and debiasing methods on predictions of novice performance. *Journal of Experimental Psychology: Applied, 5*, 205–221.

Hoch, S. (1987). Perceived consensus and predictive accuracy: The pros and cons of projection. *Journal of Personality and Social Psychology, 53*, 221–234.

Kelley, C. (1999). Subjective experience as a basis of "objective" judgments: Effects of past experience on judgments of difficulty. In D. Gopher & A. Koriat (Eds.), *Attention and performance XVII* (pp. 515–536). Cambridge, MA: MIT Press.

Newton, L. (1990). *Overconfidence in the communication of intent: Heard and unheard melodies.* Unpublished doctoral dissertation, Stanford University, Stanford, CA.

Nickerson, R. (1999). How we know—and sometimes misjudge—what others know: Imputing one's own knowledge to others. *Psychological Bulletin, 125*, 737–759.

O'Mahony, J. (1984). Knowing others through the self—Influence of self-perception on perception of others: A review. *Current Psychological Research and Reviews, 3*(4), 48–62.

Page, B., & Jones, C. (1979). Reciprocal effects of policy preferences, party loyalties and the vote. *American Political Science Review, 73*, 1071–1089.

Piaget, J. (1962). Comments [Addendum to L. Vygotsky, *Thought and language* (E. Hanfmann & G. Vakar, Ed. & Trans.)]. Cambridge, MA: MIT Press.

Tversky, A., & Kahneman, D. (1974). Judgment under uncertainty: Heuristics and biases. *Science, 185*, 1124–1131.

Critical Thinking Questions

1. Nickerson suggests that we have a default assumption about what other people know. Wouldn't it make sense in many situations to assume that they know little or nothing, so as to ensure that we don't miscommunicate? What would communication be like if you assumed that the person with whom you spoke had little or no knowledge? In what situations is it *strategically* important to make assumptions about what others know: business? Athletics? Politics?

2. Nickerson assumes that we hypothesize that a "random other person" knows what we know. Do you think that most people would assume that they are like a random other person or that they are a bit smarter than a random other person? If they assume they are a bit smarter than a random other person, why aren't they reducing their estimate of what others know to account for that fact?

3. The author describes a number of disadvantages to assuming that others know what we know; these problems occur because we believe that others have more knowledge than they actually do. What would be an alternative way of assessing how much knowledge people have? Suppose this alternate method led one to assume that people have *less* knowledge than they actually do—would that be better or worse?